D0419058

# THE BRUNEVAL RAID

## Flashpoint of the Radar War

BY THE SAME AUTHOR

*Maquis*

*Horned Pigeon*

*My Past Was an Evil River*

*Isabel and the Sea*

*Through the Unicorn Gates*

*A White Boat from England*

*Siesta*

*Orellana*

*Oyster River*

*Horseman:*
Memoirs of Captain J. H. Marshall,
edited by George Millar

George Millar

# THE BRUNEVAL RAID

## Flashpoint of the Radar War

With a Foreword by Admiral of the Fleet
The Earl Mountbatten of Burma

THE BODLEY HEAD
LONDON SYDNEY
TORONTO

*To Georges Molle,*
*Mayor of Vieilley*


ISBN 0 370 10268 1
Printed and bound in Great Britain for
The Bodley Head Ltd
9 Bow Street, London WC2E 7AL
by Redwood Burn Limited
Trowbridge & Esher
Set in Monotype Bembo
*First published 1974*
*Reprinted 1974*

# *Contents*

# *Illustrations*

*List of plates*



*Maps*
*(drawn by Maureen Verity)*

# *Acknowledgments*

We happened to have the honour to be the very first British soldiers, Rifleman Jones, Rifleman Skinner, and Second Lieutenant Millar, whom Rommel's advance guard brushed up against (some brush!) when (see *Prologue*) the German general made his final spectacular advance on Egypt. After a few ups and downs and whirligigs we were netted by those excellent troops, and I spent the following twenty months in the bag.

Accordingly I heard and knew nothing of the Bruneval Raid until one day in 1970, hunting another fox in the rich literary coverts of the Air Historical Branch, Ministry of Defence, London, coverts then keepered by the celebrated L. A. Jackets, I stumbled upon the Raid, and finally changed foxes. First, therefore, I should like to thank Mr Jackets and his staff for their initial and continuing help, and Group Captain E. B. Haslam, who now runs the Branch.

Following an initial and invigorating research success (every conscientious writer knows how such matters proceed) I ran into a period of patchy scent and awkward terrain. But then I had, at a time of near-despair, the bright notion of writing to Earl Mountbatten of Burma, and suddenly hounds gave tongue, and we were off again at a steady hunting canter over a good bit of the vale. During this second hunt I had important assistance from Rear-Admiral P. N. Buckley, head of the Naval Historical Branch, and his staff, who were even able to turn up apposite German documents.

It seems to me that the general public, even the general reader, if the latter may be presumed to exist, little understands what a power-house of knowledge and interest is preserved in our service and civilian archives, or begins to appreciate the enthusiastic work of the archivists, who are the finest in a world where competition is keen. In this context I must thank Miss Rose Coombs, Librarian of the Imperial War Museum, and her staff. Miss Coombs was one of those intrepid WAAFs who manned the Radar scanners, and she survived a direct, long-distance attack (they watched it coming straight for them, all the way from France) on her mobile station. I also thank those two inspiring places and their very sympathetic staffs, the British Museum Reading Room and the London Library. A special word of thanks to the Rev. L. W. G. Hudson and to Tom Delmer for their expert advice.

I soon perceived that existing accounts of the Bruneval Raid contained discrepancies and vulgarities. These arose from wartime circumstance. Propaganda is a distorter. The first accounts were, quite understandably, expanded by newspapermen, and then emasculated by censorship. To reconstruct in some measure, I decided to hunt down every survivor. The bloody action of the Second Battalion The Parachute Regiment in Tunisia, when encircling Germans three times demanded surrender and three times the battalion, led by John Frost, shot and carved its way out, and then the carnage of Arnhem, had sadly depleted the ranks and numbed the memories. But the raiders had been in their twenties in 1942, and survivors there are. Some remembered. And Frost and Cox had written down their impressions soon after the Raid. Frost's account runs to twenty foolscap pages, and Cox's is little shorter.

Let me say then how grateful I am to Jock Company and to those who flew it and flew with it on the job, and how much I admire them for that and their subsequent operations in the war. I must especially thank for their help and their excellent company Major-General J. D. Frost, who now farms in Hampshire; John Ross, a lawyer in Dundee, who remembers with a clear mind and who had the prescience to keep intact his copy of Operational Orders; Charles Cox, whom I found running his prosperous wireless and television business in Wisbech, and as bright, cheerful, and efficient as even an East Anglian can be; Dennis Vernon, still an engineer but now very much in civvy street, and prospering without any fuss or bother in peace as in war; and 'Private Newman', no more like a private soldier now than then, but the most fascinating of companions; and all the others I met.

Here I should point out that nearly all the characters in this book, scientists, soldiers, sailors, and airmen, were decorated for unselfish service to their country. It seemed to me that so many letter names had to be included—such as TRE and SOE, not to mention RAF—that I simply could not include decorations. Will the holders please understand?

Lord Mountbatten opened a second door for me. Rémy is a personal friend of his. And Rémy as a subject is any writer's dream. Before I began to work on this book I had read all his published work except *Bruneval, coup de croc*. I here express my deepest gratitude to Colonel Gilbert Renault, and to those who worked with him; and most directly to the many who died for us, like Pol, boldly before the muzzles of enemy rifles, or like Bob, in agony and in loneliness.

Of course the Bruneval Raid made me ask, 'Why?' Previously I had no conception that both Germany and England discovered Radar at about the same time, and that they used it initially in different ways; and that this usage had such a vastly important effect on the outcome of the war. I prefer not to embarrass by naming them the scientists and workers of TRE (I live in Dorset, quite near Worth Matravers) whom I have consulted. Let me thank publicly only A. P. Rowe, whose modesty and apparent mildness might be thought to conceal his great services to us all. Derek Wood and Derek Dempster, in their fascinating study of the Battle of Britain, say this of Rowe (p. 138): 'In August, 1938, Watson-Watt was promoted to the Air Ministry as Director of Communications, and A. P. Rowe took his place as Superintendent at Bawdsey. Although not professing to be an electronics expert, Rowe was a first class organiser with an unusual flair for analysing problems and finding the right people to answer them.'

I feel that while I was enjoying the war with a gun in my hand those scientists were winning the war, an unshouted victory. Apart from Rowe, I will express my gratitude only to one scientist individually, and to him really belongs the central point of meaning of the book. Professor R. V. Jones has the Chair of Natural Philosophy at Aberdeen University. He is today an intensely virile and active person. He remains (as I feel he should) something of a mystery. He will probably think that I have over-dramatised his secret rôle in those vital and terrible episodes from our past.

Needless to say, NONE of the views expressed in these pages should or can be attributed to anyone but myself. Perhaps some of the more controversial passages concern the late Sidney Cotton. Soon after the war I met him and he told me some of his story; and he fascinated me. I wish there were more Australians like him in my country. And I wish Cotton were alive to read chapters 12 and 14.

I would particularly like to thank the brilliant Wood and Dempster partnership, Ronald Clark, and Alfred Price (a regular RAF officer).

The latter's remarkable study of the Radar war is particularly worth reading for its evocation of the last terrible stages, as seen through both German and British eyes.

I am most grateful to the Controller of H.M. Stationery Office for permission to quote a letter; to Mrs. Thelma Cotton, Ralph Barker and Messrs. Chatto & Windus for the numerous passages from *Aviator Extraordinary* in the two chapters about Sidney Cotton; to Constance Babington Smith and Messrs. A. D. Peters for extracts from *Evidence in Camera*; to A. P. Rowe for long extracts from *One Story of Radar*; and to Rémy for his carte blanche authority to raid the riches of his own literary production.

Finally I express my gratitude to Major G. Norton, Curator, and Colour Sergeant T. Fitch, Custodian, of the Airborne Forces Museum at Aldershot.

GEORGE MILLAR
Sydling Court
Dorset *1973*

# *Foreword*

## BY ADMIRAL OF THE FLEET THE EARL MOUNTBATTEN OF BURMA, KG, PC, GCB, OM, GCSI, GCIE, GCVO, DSO, FRS.

On the night of February 27/28, 1942, we in Combined Operations pulled off a small but completely successful raid under the noses of the Germans. In the current climate of depression this was hailed as a counter-blow which proved that the British still had their tails up. And so we had.

It was the first successful raid using parachute troops, who were dropped by the RAF on the cliff-top at Bruneval to the north-east of Le Havre. They seized all the vital elements of a Radar station and brought them safely back across a beach held by commando-trained soldiers to our assault landing craft, which were escorted by the Royal Navy and the Free French Navy.

We had excellent intelligence, not only from aerial photographs taken by the RAF, but also from the French Resistance network under their doughty chief known as Rémy, who has himself written an excellent account of the raid in a book published in France.

Not long ago, out of the blue, I received a letter from General Student, who had been watching the German version of the television series of *My Life and Times*. He wrote: 'I was particularly impressed by the suggestion you made at the beginning of 1942 as Chief of Combined Operations, namely to take the Bruneval Station in a "coup de main" from the air. This was a grand plan, just to my liking as the creator of the German paratroopers. . . . The successful execution by Major Frost sent a great shock through Hitler's headquarters.'

This raid had two quite separate consequences; primarily its importance to the war was to enable our scientists to evaluate a certain German Radar set, and secondly, its effect on morale was out of all proportion to the size of the raid. I feel, therefore, that it is high time that a book about Bruneval should be published in English, and I congratulate George Millar.

He has written a fascinating account, ranging over the whole preliminaries and explaining the real importance of the Bruneval Raid, to which I am delighted to write this Foreword. It also gives me a further opportunity to pay a tribute to the brilliant planners, and to the gallant men who made such a success of the raid.

Mountbatten of Burma
A.F.

# Prologue
# A Drop in the Ocean

*January, 1942*

Lord Louis Mountbatten, the newly appointed naval Commodore leading Combined Operations, on January 21, 1942, submitted to the three British Chiefs of Staff a proposal to raid a cliff site on the German-occupied French coast between Le Havre and Etretat.

Admiral Pound, Air Marshal Portal, and General Brooke discussed the proposal with an understandable lack of enthusiasm. They were informed that the Prime Minister was keen on the raid, having been interested in it by his scientific *éminence grise* Professor Lindemann (recently raised to the peerage as Lord Cherwell). Behind Cherwell was that young Dr Jones on the Air Staff. The object of the raid appeared to be a long-term one that might benefit the Air Force. Portal was friendly to the proposal, Pound was negative, and Brooke, as so often happened, took the middle ground, trying to assess unemotionally the possibilities and advantages of success.

At best it would be the smallest of successes in an ocean of calamity.

On that same day in January, 1942, General Rommel counter-attacked massively from the line of the El Agheila–Marada road in Libya. The leading British troops, a few Jock Columns and armoured-car units, had reached that line only nine days earlier after an advance hailed by Allied propaganda as spectacular. Few people had expected Rommel to strike back so savagely and so soon. Few people. But Pound, Portal, and Brooke had the key to the Middle East war always in mind. The key was supply. And a month earlier the Japanese had attacked Pearl Harbour, the Philippines, British Malaya, and Hong Kong. At every point Japanese arms were triumphant. The battleship *Prince of Wales* and the battle cruiser *Repulse*, both sent on Churchill's insistence to bolster the defences of Singapore, were sunk off Malaya by torpedo bombers. On December 18 the lovely island and town of Penang, with its half-oriental half-Georgian arcades, its planters' club, its ginger beer, its Malay–Chinese water villages, fell to the ferocious, swift-padding Nippon soldiers. That same day Italian frogmen holed with limpet mines the battleships *Queen Elizabeth* and *Valiant* inside the defences of Alexandria Harbour. Within the month the British and their new American ally had lost the almost unimaginable total (by earlier standards) of fifteen battleships. The day after the Italian success at Alexandria 'K' Force, the naval unit in the Malta base, dashed out to attack an Axis convoy and got trapped in a minefield. Meanwhile, on the Russian front the fast-moving German attacks slowed as the weather clamped down, hampering the U-boats and the Luftwaffe. Hitler accordingly ordered a partial withdrawal of those weapons from Russia, and an all-out attack on Malta. One of the newly arrived U-boats sank a British cruiser. The Royal Navy had (temporarily) all but vanished from the Mediterranean basin. Malta stoically endured the rain of bombs. They did not even matter as much as usual because Grand Harbour and the Sliema Creeks were empty.

Now, while British convoys to the Western Desert round the Cape of Good Hope were being bled by an increasing number of U-boats and by the drain to the Far East theatre of war, German supplies passed across to Tripoli unimpeded. These included large numbers of the new Mark IV tank with its 75-mm gun, a war vehicle that would complement the excellence of Rommel's tactics, his refusal to waste steam in small forays, his steady speed over any sort of terrain, his magnetism for the soldiers, German and British. And ahead of Rommel, the Chiefs of Staff knew, lay Egypt, the Suez Canal, Palestine—suddenly attainable.

To them the war situation was a prolonged nightmare. . . . Should the Japanese advance west across the Bay of Bengal they could cut the Canal Zone's lifeline up the Red Sea. Meanwhile, apart from the direct threat now offered by Rommel, the Germans menaced the Canal by land encirclement. During four months of 1941 their armies had bitten their way five hundred miles into Russia. (They claimed—exaggeratedly—to have killed ten million Russians.) They had taken Kharkov and Kiev, had encircled Leningrad, and were poised on the outer fringes of Moscow when winter, the Russian winter, closed his freezing jaws. Another comparable German advance in the spring and summer of 1942, and they would hold the Caucasus and the main Russian oil wells. Then they would coast down from the north and threaten the whole Middle East.

Yet was there a change, the suspicion of the beginning of a change?

It *appeared* that the Germans had fought to their limit for five days to storm Moscow while the snow fell thickly in the short daylight hours and in the prolonged hours of bitter night. It *seemed* that they had failed, and on the very day that the Japanese had bombed Pearl Harbour. So fully did Churchill believe that 'Zhukov was attacking the Hun strongly and flinging him back', so warmly did he welcome America's entry into the war against Japan and Germany, that that night he had gone to bed 'saturated and satiated with emotion' to enjoy the sleep of 'the saved and thankful'. The three Chiefs of Staff were invigorated by Churchill's optimism. But it was their task to make his projects workable, and in strategic matters they had to pass his optimism through the filters of practicality.[1]

At sea England's plight would have looked irredeemable had not the world, including Germany, been obsessed with the worth and valour of the Royal Navy, and had not the Navy carried out its nearly impossible duties in its own cheerful and professional way. When the combined strength of the Wehrmacht and the Luftwaffe had knocked out Poland, Denmark, Norway, France, and the Low Countries, and the Royal Air Force had seen the Luftwaffe off until a twentieth-century repeat of the Norman Conquest seemed too much of a gamble, Hitler had decided to starve Britain out. The French Biscay harbours had been turned into submarine bases. And now, in January, 1942, Admiral Dönitz had five Atlantic U-boats operating for every one he had had only eighteen months previously. The graph of their sinkings showed an impressive rise. It had reached four hundred thousand tons a month, and would go much higher with the entry of the United States. The Germans were having no difficulty in producing more and better submarines (but there was a parallel with the Royal Air Force in the Battle of Britain; the shortages of trained and seasoned crews were making themselves felt, even when the Atlantic battle was being won).

And again the three very senior officers knew of a secret ray of hope. At Worth Matravers, on the chalk downs of the Dorset coast, it was said with confidence that the scientists of TRE (Telecommunications Research Establishment) had developed a gadget that, in the aircraft of Coastal Command, would beat the U-boats to hell. One could believe almost anything of those fellows, an unorthodox bunch. At Swanage, near Worth, they ran weekly discussions, arranged for senior and junior serving officers, politicians, big hats from industry. They called these meetings Sunday Soviets. People like Tedder and the

Prof (Cherwell) said they were wonderfully stimulating. Apparently a portrait of Göring hung in the room where the Soviets were held. Very rum.

Apart from the U-boats, the war at sea contained another major headache, that of the capital ships. Churchill had been born into the battleship era. He could never get battleships out of his head, though many of his subordinates insisted that a battleship was about as war-worthy as a zeppelin. Although the Germans had lost the *Graf Spee*, scuttled in the shallows of the River Plate after a failure in action against three 'weak' cruisers which the pocket battleship should have annihilated, and the *Bismarck*, which had performed according to the book, taking HMS *Hood* down with it, they had commissioned the *Tirpitz*. Brand new, and said to be the world's most potent warship, *Tirpitz* threatened to break out at any moment from her North Sea bases and wreak havoc. Meanwhile at the other (western) end of the Channel three formidable German ships, the battle-cruisers *Scharnhorst* and *Gneisenau* and the cruiser *Prinz Eugen*, were embayed, growling, in Brest. With such potential marauders on either flank, the Royal Navy had to keep a powerful counter force on standby in the central position, Scapa Flow. Britain was getting desperately short of marine tonnage. The anti-submarine invention in Dorset would have to be damned good, and its introduction to Coastal Command must not be long delayed.

As for the civilians, every part of the British economy was regimented. The country would not recover from the moral effects of rationing, of 'fair shares' for a quarter of a century and more, long, long after the end of the war. As Britain caught up Germany as a manufactory of weapons, rationing went far beyond the limits of food and clothes and fuel. For example, on January 1, 1942, the Board of Trade produced standard gold wedding rings of nine carats, to be sold at a price of one guinea. On the twelfth of that month the Ministry of Food was particularly active. Fat and sugar rations were reduced to eight ounces of sugar and six ounces of butter and margarine (not more than two ounces of butter); those were weekly rations per adult person. The amount of milk issued in schools was reduced to one third of a pint per pupil per day. Poultry and rabbit wholesalers were warned to take out licences. The Ministry engaged in an advertising campaign to push 'National Rose-Hip Syrup, rich in Vitamin-C'. But perhaps the Ministry's most characteristic offer was this: milk powder damaged in transit would be made available for British cats 'employed in vermin destruction' in warehouses containing over twenty-five tons of food.

German intelligence, receiving the English newspapers and magazines via Lisbon and Geneva and monitoring the BBC programmes, pondered such legislation and deduced that the once-fat enemy was starving and must surrender when a further hundred thousand tons of her shipping had gone down. In a similar frame of mind the Ministry of Economic Warfare and other British sources maintained that the German people under the Nazis guns-before-butter programme had been on short commons before the war, and were now in a worse case. They were wrong. Germany had conquered countries with splendid and varied agricultural productions. The conqueror drew heavily on them for supplies, to the detriment of the local consumers. Rationing was less strict than in England, and the Germans were the more inclined to black market dealings. Then, there were so many foreign workers and working prisoners-of-war in Germany that labour shortages had not appeared as they had in Britain.

That the inhabitants of the British Isles accepted their régime without much grumbling was a sign that they meant to win the war. They seemed to have gained from the war, to have drunk of the elixir of life. They also seem, at any rate in retrospect, to have been a little mad. With the war, classical music had become more admired, and remains so to this day. But the popular songs, when the war had reached this razor's-edge stage were, *I've Got Sixpence*, *Amapola*, *I-I-I-I-I-I Like You Very Much*, *Kiss the Boys Goodbye*, *Bless 'em All*, *Yours*, and *The White Cliffs of Dover*. *Lili Marlene* had filtered home from the desert and had been given an English translation.[2] It was still the age of The Great Band, those enormous bodies of musicians who contrived to make so little noise. The factories churned on solaced by thick tea and *Music While You Work* played by such loved combinations as those of Jack Payne, Harry Roy, Geraldo, Joe Loss, Victor Sylvester, Billy Cotton, Mantovani, Jack Jackson, and Henry Hall. Dancing was a national pastime, if one could call it dancing, and at night in millions of houses the wireless set held sway. Yet with all that apparent inanity there was hardly an infant in the land, let alone a parent, who did not realise that much of the food, and all the petrol, had to be brought to Britain in ships whose seamen, in the night as in the day, faced and often found a horrible death.

To make life even less easy than it would otherwise have been for the Chiefs of Staff, the Prime Minister had suddenly come home by air. Five days after Pearl Harbour, he could no longer keep away from Washington. He had left Glasgow in the battleship *Duke of York*, the twin of the unfortunate *Prince of Wales*. His deputy, Major Attlee, and the new Chief of the Imperial General Staff, General Sir Alan Brooke, were in charge of the day-by-day running of the war from London. These two were more than capable of the task. But one of them at least immediately felt that the Prime Minister's visit to America might have dangerous effects.[3] The Americans, new to the war, did not realise how fully stretched their ally was, nor how unprepared they were themselves. President Roosevelt had much faith in General Marshall, who devised an Anglo-American defence against the Japanese that, initially, could have little basis in genuine strategy. The association (ABDA) apart from bolstering China and relieving General MacArthur in the Philippines, was to defend a vast area stretching for four thousand miles from the Sumatra Straits and the entrance to the Indian Ocean to the Coral Sea and Australia. The Japanese were to be denied what they were determined to get, the oil and rubber of Borneo, Java and Sumatra; they were to be denied the Singapore base and the Burma access to India and China. Churchill, for all his natural love of the grandiose in military affairs, saw that the plan must be out of touch with reality. But as always, he put first things first. In his view, and who can say today that he was wrong, the main objective was to build, cement, and enlarge the American alliance. His own closest military advisers, Alan Brooke in London and Sir John Dill, who was with him in Washington, had tried to impress upon him the folly of reinforcing weakness in Malaya, Singapore, and Indonesia, and the pressing need instead to reinforce Burma as the bastion for the defence of India. On the day after the *Prince of Wales* and the *Repulse* were lost the War Cabinet had agreed to transfer Burma from the Malayan to the Indian Command, and to re-route the Eighteenth Division, which was at sea on its way to the Western Desert—to send it from the Cape direct to Bombay. In Washington this decision was rescinded in Brooke's absence and despite Dill's remonstrances.[4] The fresh division which would have been invaluable in the Middle East, but now was deemed essential for the defence of India, was sent to its doom in Singapore.

That fate, on January 21, 1942, was still twenty-five days distant. But Brooke, Portal, and Pound knew it would come, and its inevitability horrified them and plagued their hearts and consciences.

So . . . What about this raid proposed by Mountbatten? Its technical reason was top secret. British scientific intelligence had very recently discovered that Radar, which had been regarded as a British trump card and which had played a major part in the victory of the Battle of Britain, was understood and employed by the enemy. Much of Britain's war production and effort was being channelled into Bomber Command, whose increasingly heavy attacks on Germany were meeting an increasingly effective defence. Losses at this time were running at four heavy bombers out of every hundred sent over German or German-occupied territory, often more, seldom less. In the light of the most recent assessments these losses were attributable to a disquieting factor—a question mark in fact—German Radar.

With the help of the Photographic Reconnaissance Unit (PRU), that exclusive high-speed Spitfire unit of the Royal Air Force, young Dr R. V. Jones and his helpers in Air Intelligence had quickly come to an understanding of what a layman might describe as the long-range, non-precision German Radar unit, the *Freya*. They had the *Freya* taped. They could locate it, listen to it, decode with ease its messages sent back to German guns and fighters, and at any moment, if they wanted to, they could jam it, neutralise it or, better still, feed it false information. So far so good.

But they knew of the existence of another German Radar unit, one with a shorter range but more precision; a unit that (it was feared) could clamp on to a Lancaster or a Stirling bomber and, holding it in its narrow, invisible beam could set the killers—German night fighters—on to it. This unit was small, and presumably easy to manufacture in quantity. It was taking the most costly of British lives, RAF lives. Many such units had been located at long range by 'smeller' reconnaissance Wellingtons of the RAF. Now at last, thanks to a brilliant bit of work by a Spitfire pilot, one of those infernal machines (apparently it looked like an enlarged electric bowl fire) had been located in an attackable site on the Channel coastline of France. It was on a cliff-top immediately north of a small village called Bruneval, itself some twelve miles north of Le Havre. Dr Jones and the scientists of TRE urgently wanted the vitals of that German Radar unit so that they could devise means of beating it, and Mountbatten, on behalf of Combined Operations, stated that the vitals could be snatched.

He only asked for minimal forces to do it: one company of the Parachute Regiment; one section of airborne Royal Engineers; a couple of Radar mechanics from the RAF establishment; a squadron of Whitley bombers to take the parachutists to their objective, and suitable light naval forces to bring them back to England from it.

Bruneval, the three Chiefs of Staff saw, was in a ravine that ended in a cliff-girt beach guarded by concrete pillboxes. An evacuation of a small force by sea in perfect weather seemed feasible. The PRU cover was excellent, and local intelligence was being sought through de Gaulle's lot, who were confident that they could get it.

No time was lost in approving the Bruneval project. Contact was made immediately, through the leader of the airborne forces, Major-General F. A. M. Browning[5], with the Second Parachute Battalion, which was in process of forming at Hardwick Hall, near Chesterfield.

NOTES TO PROLOGUE

1. Bryant, *Turn of the Tide*, p. 282.
2. Longmate, *How We Lived Then*, p. 416.
3. Bryant, *op. cit.*, pp. 290–5.
4. Bryant, *op. cit.*, p. 295.
5. Daphne du Maurier's husband.

I

# Jock Company

*January, 1942*

John Frost, a regular soldier commissioned in the Cameronians with a conventional background (father a Brigadier in the Indian Army; educated at Wellington and Sandhurst; a shooter and a foxhunter) was adjutant of the Second Parachute Battalion when his commanding officer was required to send one company down to Salisbury Plain.

'C' Company, the most efficient at Hardwick Hall, was known as Jock Company, being made up almost entirely from Scots regiments. But at that time it was led by an Englishman from an English regiment. Frost was told that if he could within the space of a week complete his parachute jumps to the statutory number (he had so far only done two, and had got water on the knee as a result) he, as a Cameronian, might change places with the Englishman, Major Philip Teichman. Meanwhile, though, Teichman was sent with the advance party to the Plain, on the understanding that if Frost, through accident or bad weather, failed to complete his jumps within the week Teichman would keep the command, which he badly wanted. So did Frost.

He had just passed his thirtieth birthday. He was tall, healthy, highly strung, with a secretive look to him. When he arrived at Ringway, the efficient and well-informed parachute training centre near Manchester run by the Royal Air Force, he saw that the staff took a special interest in him. He even heard them talking about 'the last time', a reference, he was sure, to the only previous British operation with parachutists. A band of them led by 'Tag' Pritchard of the Royal Welsh Fusiliers had been dropped to destroy an aqueduct in Southern Italy. Not a single parachutist had returned. They were said to be languishing 'in the bag' in Italy, which could not be much fun.[1]

Fog, strong winds, and a shortage of aircraft at Ringway made his qualifying jumps a matter of luck. He completed them within hours of the time-limit, hurried back to Hardwick Hall, took command of

'C' Company, and entrained for Salisbury Plain. Teichman was anything but pleased to see Frost, and he said so. Both had become parachutists to find action, and Frost, perfectly understanding the other's anger, felt that it was a bad start.

His company had been assigned quarters in part of Tilshead Camp. The camp was run by the Glider Pilots' Regiment, which was forming, and expanding there. Frost thought Tilshead a miserable hole, and he was not cheered by the news that General Browning, who commanded the First Airborne Division from his headquarters at Syrancote House, near Tilshead, intended to inspect 'C' Company the following day. 'Boy' Browning had been a disciplinarian adjutant at Sandhurst, and he could be relied on, Frost knew, to notice every deficiency in turn-out and conduct. 'Our men were a wild crew,' he says.[2] 'At that stage of the war clothing and equipment were scarce, and for a few months we had been concentrating on toughness and on weapon- and parachute-training. We'd had little time for drill, and still less for making ourselves look glamorous, or even clean. After a prolonged and uncomfortable railway journey, the Jocks had found time to work the dreadful Tilshead mud deep into the fabric of their uniforms. They looked horrible.'

When Browning had inspected the company he led Frost aside. 'Just let Peter Bromley-Martin know exactly what you need in the way of transport, stores, and equipment. And see here, Frost! Every man is to get a new uniform, for that is the filthiest company I ever saw in my life.' Later that first day Browning's liaison officer, Bromley-Martin, Grenadier Guards, arrived in the company office and was introduced by Frost to his young officers. They had looked forward to meeting him. His report on his first parachute jump (on February 4, 1941) was regarded as a classic.

He had jumped fourth, following his friend, H. O. Wright. 'The next recollection I have,' he wrote,[3] 'is of Major Wright with parachute open and canopy fully filled, some one hundred and fifty feet *above* me. My parachute, sir, had not then fully opened, and I had the gravest doubts as to whether it would function before it had been repacked. I was unable to devise a method of repacking it in the limited time at my disposal. As I was also unable to think of any satisfactory means of assisting the contraption to perform the functions which I had been led to suppose were automatic, in my submission I had no alternative but to fall earthwards at, I believe, the rate of thirty-two feet per second, accelerating to the maximum speed of one hundred and seventy-six feet per second. . . . This I did. . . . And having dropped a certain distance, my parachute suddenly opened, and I made a very light landing.'

Bromley-Martin told them that they had been brought south to stage an exercise judged to be vital for the future of the Airborne Division. Churchill was known to prefer the notion of sea-borne landings on the Axis perimeters, using commandos. He was going to take a lot of convincing as to the usefulness of parachutists. 'C' Company was to simulate a raid on an enemy headquarters in occupied territory. The demonstration would be on the Isle of Wight. The War Cabinet and Churchill would be present. Alton Priors, near Tilshead, would be their initial training ground for moving from the dropping zone (DZ) to their objective, and from the objective to the imaginary coast. They would be issued with the most modern and most lethal weapons, and their comfort would be seen to. At a later stage in their training for the demonstration they would work with the distinguished RAF unit that would drop them, and with the Royal Navy landing craft that would evacuate them, the simulated operation successfully completed. For the exercise, Bromley-Martin concluded, the company would be split into assault parties of differing sizes, each trained, and equipped, for a specific task. The tactics, in short, would be laid down by headquarters. Frost violently disagreed with that, particularly for a night operation, and he said so. Bromley-Martin departed from a hostile atmosphere.

Frost and his platoon commanders were furiously disappointed. They were tired. They were depressed by the mud and the rain, and even by their hosts, the glider pilots. The staff officer's briefing had seemed a piece of unctuous nonsense.

Next morning Bromley-Martin reappeared in the company office, and told Frost that he wanted a private interview. After emphasising that Frost must maintain the 'Prime Minister cover story' with his officers and his men, and must ensure that they really believed it, Bromley-Martin went on, 'You'll be taking your company over to France before February is out.[4] Now it's up to you to see that everything works properly and that your Jocks are so fit they're jumping right out of their skins, or you won't have a hope of bringing them out alive.' The raid, he said, had a special objective which he was not yet at liberty to describe. And the enemy's dispositions were complex. Frost must accept it from him that, in order to secure the line of retreat, or rather withdrawal, and also to give maximum protection to the men dismantling the objective, the plan of battle would be drawn up for him by headquarters. If he wanted to lead the raid he must fall in with those terms, even if privately he disagreed with them.

'Let's have a gin,' Frost said, determined to accept, although he felt that he was being blackmailed. Division's insistence on a rigid plan continued to worry him (and was to infuriate him during the raid).

'You'll have noted my use of the verb dismantle,' the staff officer said. 'That's why you've found under your orders here a section of the First Parachute Field Squadron, Royal Engineers. Under the present plan for the raid, which may of course be altered, four of your sappers will carry out an anti-tank role. The rest of them will be in your dismantling party.'

'Sounds exciting,' Frost said drily.

'Also in your dismantling party will be two RAF sergeants. Then, except for one man, Newman I believe they call him, your strength will be complete. . . . I don't think I should tell you about Newman yet. . . . I must say I find the whole thing fascinating.'

'That's good,' Frost said. He had the impression that Bromley-Martin was planning the raid, and he did not like that idea at all.

## NOTES TO CHAPTER I

1. The captured parachutists were in Gavi POW Punishment Camp with the author. They were an unusually impressive batch of soldiers. Their mission, to blow up an aqueduct in Southern Italy, had been hastily conceived. They had accomplished it with ingenuity, courage, and humanity. . . . The Bruneval project was in fact only the second parachute operation carried out by the British Army.
2. Frost's written account.
3. *By Air to Battle*, p. 9.
4. Frost's written account.

# 2

# The Attics of Science

*1934*

'Radar' is a reversible word brought to the war when the Americans with their talent for new words came in and helped to build on the framework developed by British scientists, service departments, the radio industry and the General Post Office. Radar is an abbreviation for Radio Direction and Ranging. It is a good word, and will be used here, although in the 1930s when England and Germany independently discovered Radar (each refusing to believe that the other might have it) they had their own cover names for it, Radio Direction Finding or RDF in England, Dezimeter Telegraphie or D/T in Germany.

What is scientific discovery? Occasionally it is an abrupt and wonderful breakthrough; usually it grows like a tree, one research branch topping its predecessors until suddenly the tree is powerful. Radar grew from the German physicist Heinrich Hertz, born 1857, died 1894. Hertz, who himself continued the work of Faraday and Clerk Maxwell, may be said to have discovered and demonstrated the true nature of radio waves. He showed, for example, that the waves were reflected from metal sheets. But he could see no practical use for this discovery. Three years after Hertz died another German, Professor Ferdinand Braun, invented the Braun tube, which the Germans continued to call by that name, and the English called the cathode-ray tube (it is the television screen). In 1904 Professor Ambrose Fleming working at University College, London, made the first true radio valve, the diode. That year, too, a young German, Christian Hülsmeyer took out British Patent No. 13,170 for his invention, a 'Hertzian Wave Projecting and Receiving Apparatus'.[1] His patent comprised a radio transmitter and a receiver mounted side-by-side in a ship. He claimed that the waves sent out by his transmitter if they encountered a metal body on the sea, presumably another ship, would close an electrical contact in the receiver and ring a bell.

How much nearer to Radar could you get than that? But Hülsmeyer's idea was unpopular at a time when the British Admiralty and its friends and rivals were thinking in terms of armour and gunnery. Three years later, in 1907, an American, Professor Lee de Forest, improved on Fleming's valve. In 1924 two Americans, Dr Gregory Breit and Dr Merle A. Tuve evolved a technique for sending out a series of radio 'shouts', or short pulses; and using their technique Professor E. V. Appleton in England was able to calculate the height of the Heaviside layer, the layer of radio-reflective ionised gases surrounding the earth (some sixty miles). Five years later, in 1929, Professor Hidetsugu Yagi published in Japan the results of his successful experiments with directional aerials which, he proved, made it possible to send out radio signals in narrow beams.

So, by the 1930s, the scientific groundwork existed for the discovery of Radar. How many more such discoveries, one wonders, lie among the lumber in the attics of science, waiting for the kiss of recognition, of life—or for the breath of war?

* * *

Dr Rudolf Kuhnold, a sonar expert working with the German Navy, independently researched with radio waves (he had not heard of Hülsmeyer); and in 1933, using new valves developed by Philips in Holland, he set up experimental transmitting—receiving apparatus on a balcony overlooking Kiel Harbour, aimed his dished aerials at a battleship five hundred metres away, and got satisfactory echoes in his receiver.[2]

A company, GEMA, was formed to develop Kuhnold's discovery and to sell it in the growing German market for work relating to armaments. In October, 1934, demonstrating an improved version of his apparatus to Navy staff and engineer officers at Pelzerhaken, he got strong echoes from a ship at a range of seven miles. GEMA was given a research grant of seventy thousand Reichmarks. Turning from continuous waves to pulses, Kuhnold by 1936 had increased the range of his naval set to twelve miles. This success was reported to Hitler and all Chiefs of Staff.

In Germany the question posed by Dezimeter Telegraphie (it was put under the cover of the German Post Office) was this: How could it best be used in attack? Well, it could plainly be used for setting and ranging guns, first naval guns and later anti-aircraft ones and also, presumably, searchlights. GEMA produced for the Navy the *Seetakt*, a gun-laying ranger. The prototype was tried out at sea in 1937, and the pocket battleship *Graf Spee* was fitted with *Seetakt* when in the summer of 1938 she did her publicised Spanish Civil War standing-patrol. *Seetakt* was not publicised, needless to say.

Meanwhile Dr Kuhnold's GEMA company had also produced (in 1936) a promising early-warning Radar with a rectangular aerial revolving round a vertical axis. It was semi-mobile, easy to man and to service, and its initial range on aircraft of thirty miles was soon increased to fifty and then to seventy-five miles. It was called *Freya*, and once more GEMA did business, this time with the Luftwaffe.

That year GEMA faced a powerful commercial rival when the Telefunken manufacturing complex entered the field of D/T. Telefunken knew what might sell in the German military market of that period. The firm evolved something very intelligent, and many years ahead of its time. The *Würzburg* had a round dish aerial which could follow any fast-moving target (fast-moving that is by aircraft standards). It had a useful range of some twenty miles, was comparatively simple to handle, and had remarkable accuracy. It was an obvious complement in any defensive scheme to the wider-ranging but less accurate *Freya*. Further, it was a workhorse, tough and durable and mounted on four wheels that were lifted when it went into action. It was therefore suited for forward duty with the Luftwaffe, working with the peerless German 88-mm flak gun. It was after seeing an early demonstration of the *Würzburg* in its anti-aircraft role that Reichsmarschall Hermann Göring dropped his celebrated brick when he announced that the Ruhr would never be bombed. If he was so impressed by the *Würzburg*, and nobody could deny that its performance was phenomenal, why did he not take steps to accelerate its production? Why? Because it was not aggressive. The then German philosophy was attack, strike, annihilate.

If the *Würzburg* suffered initially from neglect and consequent production difficulties, and if it never quite fulfilled its early promise as a gunlaying instrument, that was because few experts on either side in the Second World War understood that the flak gun was as clumsy a weapon as the bomber. But when the *Würzburg* worked in conjunction with the night fighter it was a different story—as we shall see.

Germany, then, had made a good start in Radar before the war. But Göring summed up the official attitude to the new science when he said of it to his commander of Luftwaffe Signals, General Wolfgang Martini, 'Radio aids contain boxes with coils; and I do not like boxes with coils.'

* * *

In England Radar may be said to have begun one summer morning in 1934. The only assistant in the office of the Directorate of Scientific Research at the Air Ministry left London for Biggin Hill aerodrome to see a sound locator working. His name was A. P. Rowe and he was twenty-six years old.

England's early-warning system then consisted of sound locators. That day at Biggin Hill the sound locator was nullified not by simulated enemy action but by a Kentish milkman whistling as he drove his cob along an aerodrome road that should have been declared out of bounds, the churns behind him jangling pleasantly. Rowe saw nothing pleasant in the scene. He hurried back to the Air Ministry and called for every file on early-warning systems for air defence. There were fifty-three files.[3] When he had gone through the lot, and found not a workable idea in any of them, he wrote a letter to his chief, H. E. Wimperis.

Wimperis in his turn wrote to Lord Londonderry, Secretary of State for Air, urging that a small committee be set up including independent scientific members to find a new method of air defence. What he was virtually saying, this Government servant, was, 'The situation is desperate. Bring in the best outside brains there are.'

Had not Lord Londonderry appointed that particular committee England might speak German today. The men chosen were Wimperis, who had suggested it, with Rowe as secretary, and three distinguished scientists, Henry Tizard, Professor A. V. Hill, and Professor P. M. S. Blackett. Tizard was England's leading defence scientist. The son of a naval officer, he served in the 1914–18 war as a gunner, then in the Flying Corps. He was an unusually good pilot, and before embarking on his scientific career he served for a while with the RAF. He had constantly, as Rector of Imperial College from 1929, been the key man on various Air Force committees, and was known for his habit of cutting through to the marrow of any problem.[4] Hill, a physiologist, had initiated operational research (co-operation between scientist and fighting man) in the First World War. Blackett had served in the Navy until 1918 and had gone then to the Cavendish at Cambridge, where he worked with the mighty Rutherford. As a physicist he was known far outside his own country and he was to play a vital part in the sea war that lay ahead. The committee had no weak link.

The first meeting was on January 28, 1935, when they came to a vital decision (the truth being that both Wimperis and Tizard had already been investigating the problem). They asked Robert Watson-Watt to give them a paper on the radio detection of aircraft.

Watson-Watt worked in the Radio Research Station, housed in a group of wooden huts at Ditton Park, near Slough. Like Professor Lindemann, Watson-Watt was a sharp-witted, sharp-tongued personage with many admirers and not a few enemies. Ten days before the first meeting of the Tizard Committee (as it was known from its inception) Wimperis had asked Watson-Watt about the possibility of a death ray which would kill air crew at long range.

That afternoon Watson-Watt talked about the death ray notion (then a popular fancy) with his assistant, A. F. Wilkins. The pair agreed that the death ray would take too much power and was a non-starter.

'Wonder what else we can do to help them?'[5] Watson-Watt said, and during the subsequent talk they recalled a GPO report. The Post Office had been experimenting at Dollis Hill with VHF radio, trying to find an economical means of communication between the Scottish mainland and the Hebrides. And the engineers had commented on a 'flutter' in their earphones each time an aircraft passed. Did this not mean that aircraft (and all the new ones were made of metal) re-radiated radio waves? And if so could not re-radiation be used for an early-warning system. Watson-Watt asked Wilkins to work out what power would be needed to get a detectable signal from an aircraft and Wilkins's calculations were favourable, suggesting that aircraft might be located at long range with radio waves or pulses.

On February 14 the Tizard Committee were handed Watson-Watt's paper entitled, *The Detection and Location of Aircraft by Radio Methods.* None of the five took long to draw the meat out of such stuff. Watson-Watt was engaged immediately in prolonged discussion which continued over luncheon at the Athenaeum. Next morning Wimperis, representing the Committee, asked Air Marshal Sir Hugh Dowding, then in charge of Research and Development, for £10,000 from public funds. Dowding insisted on a practical experiment.

Overnight Wilkins got his equipment together—an improvised receiver linked to a cathode-ray oscillograph—and stowed it in the

Radio Research Station's caravan, which was hitched to a Morris car. Next day he and Dyer, the Station's driver, drove north and parked the caravan near the old Cavalry School at Weedon in Northamptonshire. An aircraft was going to fly through the BBC's short-wave transmissions emanating from the high masts at nearby Daventry.

Squadron-Leader R. S. Blucke, commanding the Farnborough Flight, was ordered to fly the demonstration in a twin-engined Heyford bomber. He thought it odd that he was briefed by a young civilian (Rowe). February 26 dawned clear, with a strong wind from the south. Blucke flew over Daventry at the required time and altitude, one thousand feet, fired a Very light signal, turned east, climbed to six thousand, turned at a stipulated mark, and flew back to Daventry at the Heyford's maximum of 130 knots. Over Daventry, bored with the whole thing, he again fired a Very and set course for home.[6]

Below in the caravan Watson-Watt, Wilkins, and Rowe watched the oscillograph, which showed the 50-metre beam from Daventry as a straight line. But when the Heyford lumbered into the beam the line was bent. An oscillation of over an inch was registered. Rowe went back to London to report, and the next day the Treasury handed over the money.

Blucke imagined that he had been flying 'some scatty BBC stunt'. But his simple flight laid the foundation for the defence of a Britain whose initial Allies were to crumble before the power of the eagle.

## NOTES TO CHAPTER 2

1. Price, *Instruments of Darkness*, p. 56.
2. Price, *op. cit.*, p. 58.
3. Clark, *The Rise of the Boffins*, p. 28, etc.
4. See Clark's *Tizard*, also *The Rise of the Boffins*, pp. 31, 32.
5. See Watson-Watt, *Three Steps to Victory*, Clark, *The Rise of the Boffins*, p. 34, etc.
6. Clark, *op. cit.*, pp. 36, 37.

# 3
# Cox and 'Newman'

*February, 1942*

On February 1, 1942, in the Chain Home Radar station at Hartland Point in North Devon, Sergeant C. W. H. Cox was handed a railway warrant to London and ordered to catch the noon express out of Bideford. Cox, whose father was a postman and whose mother was an actress, was a cinema projectionist and radio ham from Wisbech, Cambridgeshire, a true little East Anglian, quick, staunch, perky, humorous, efficient, and patriotic. Before joining the RAF in 1940 he had never been far from Wisbech. He had never been in a ship, nor in an aeroplane. He was one of the best Radar mechanics in Britain.

In the morning he reported to Air Commodore Tait at the Air Ministry.

'You've volunteered for a dangerous job, Sergeant Cox.'

'No sir.'[1]

'What d'ye mean, no sir?'

'I never volunteered for anything, sir.'

'There must be some mistake. I asked for volunteers from among the comparatively few with exactly your qualifications. . . . But now you're here, Sergeant, *will* you volunteer?'

'Exactly what would I be letting myself in for, sir?'

'I'm not at liberty to tell you. . . . I honestly think the job offers a reasonable chance of survival. It's of great importance to the Royal Air Force. And if you're half the chap I think you are, you'll jump at it.'

'I volunteer, sir.'

Cox was promoted to Flight-Sergeant and was given another railway warrant, this time to Manchester. They also gave him a chit. He was to report to the adjutant of No. 4 PTS at Ringway.

He could not make Ringway out. 'Bus-loads of soldiers kept entering and leaving. Some had queer pots of helmets on, rather like the Boys'

Brigade.' Finally he asked the RAF Sergeant in the guard room, 'What *is* this joint, Sarge?'

'Number four PTS mate.'

'What's PTS when it's at home?'

'Parachute Training Squadron.'

Twelve days later he was told that they would let him out next morning for Tilshead, on Salisbury Plain. That night he would make his last jump. He got into his harness once more and entered a balloon basket with one of those supermen, a Sergeant-instructor, RAF. They rose silently to five hundred feet. The balloon tugged at its cable, making it grunt.

'Ready son?'

'Sarge.'

'Then let's have a good jump. . . . Chin up. Hands by your sides. Relax. Get ready. . . . Go!'

He gauged his drop by a dark-purple line of trees, made a good rolling landing, and gathered up his 'chute. He doubted if he had ever felt better in all his life.

When he reached Tilshead Major Frost and Captain Ross, the second-in-command, gave him an initiatory week of PT, route marching with full kit, unarmed combat, weapon training, knife fighting, barbed wire scaling (one parachutist lay across the wire and the others ran over him), and night patrols. He rather liked his fellow Sergeants in 'C' Company. 'If you could understand half what they said they weren't a bad bunch.'

Cox gathered from what he could understand that his new friends thought they were preparing for an exercise. He, of course, was in no doubt at all that he was destined, for the first time in his life, to make a short trip abroad. He was not surprised when Sergeant-Major Strachan told him that he would be working with the section of Royal Engineers under Lieutenant Vernon. Nor was he surprised when a mobile gun-laying Radar on loan from the local Anti-Aircraft Command was parked inside the perimeter of the camp, and he was told to explain to the REs exactly what it was and how it worked. Dennis Vernon, then twenty-four, was a Londoner who had spent much of his time at Cambridge, first at the Leys School and then reading economics at Emmanuel. Many believe the Royal Engineers to be the most impressive unit in the British Army, and Vernon was a junior officer of quality. Cox saw this at once. It had been proposed that there should be two Radar mechanics, but the second one had injured himself in his parachute training, and it was Cox who said to Air Commodore Tait that Lieutenant Vernon would be every bit as handy with any Radar as he was himself. In the 'exercise', Vernon was told, four of his sappers would be detached from the section in 'an anti-tank role'. The remaining six, himself, and Flight-Sergeant Cox would be 'dismantling the objective'. Was he familiar with the Leica camera, he was asked. (He was.) And had he taken photographs with an automatic flash, as fitted on the Leica. (He had.)

Immediately after their arrival at Tilshead Frost had promoted John Ross of the Black Watch, one of his platoon commanders, to be second-in-command and a temporary, acting, unpaid Captain. Company Sergeant-Major Strachan was also Black Watch and, Frost noted, 'the very best sort of senior NCO in the world'. Between Strachan and Ross, 'an imperturbable and extremely intelligent officer from Dundee', the administration of 'C' Company went smoothly. Those were days of parsimonious supplies for the troops at home, but not, it seemed, for 'C' Company. Ross indented one day for nine Bren-guns; the following day eighteen brand new Brens arrived. Anything he asked for was immediately supplied, uniform, boots, binoculars, compasses, pistols, torches, trucks. They got things, too, that he had not ordered, anti-tank mines and two portable mine detectors, and a new kind of sub-machine-gun, the Sten. At first acquaintance it seemed an ideal weapon for close fighting, short, light, and handy, with magazines easy to change in the dark, and a high rate of fire on 'automatic'. They soon found in training that it had been put together rapidly, and was not entirely reliable. Then they were given four No. 38 wireless sets, one for each platoon and one for Company Headquarters, and two No. 18 sets 'for contact with the Navy'.[2]

None of the parachutists thought well of the 38 set. 'It was unpleasantly temperamental and difficult to keep on net,' Frost says. 'If practice could have made the sets perfect they would have worked for us. Most nights in the early stages of training at Tilshead we were prowling about the Plain. I fear we caused the Glider Pilot Regiment some headaches. We never knew when we might appear for meals, and they were horrified at the amount of food we consumed. The men worked well, and the NCOs, taking their cue from Strachan, were first class. We worked them all just as hard as we could, but whenever we had a night in camp I made sure that those who wanted relaxation got leave to Salisbury and that transport was laid on for them.'

One day the Company, the R Es and Cox drove to Thruxton Aerodrome to meet No. 51 Squadron, which had been detailed to provide their aerial transport; traps were being cut in the floors of their Whitley bombers. Commanding 51 Squadron was one of the exceptional characters produced by the RAF during the war, Wing-Commander Pickard. Charles Pickard, tall, fair-haired, and pipe-smoking, had already seen much action, and he was a celebrity with the public because he had played (as the pilot of F for Freddie) a leading part in the RAF film, *Target for Tonight.* The parachutists instantly liked him, as everybody did whether he was hanging by his feet from a rafter in the Mess bar, proving that he could drink a pint of beer upside down, or whether he was landing a Lysander in a dark field in the Occupied Zone of France.

No sooner were they back at Tilshead from Thruxton than the Sergeant-Major brought the final addition to the strength before the Company Commander. He was a small man, but handsome, dressed as a private soldier in the Pioneer Corps. He had been detailed to join 'C' Company as German interpreter, and came from Combined Operations headquarters via Syrancote House. Frost was instructed to put him on the Company list as Private Newman. Not only did 'Newman' evidently speak perfect German, he was German. His father had come to England before the war as an enemy of the Nazis. Frost studied Newman carefully, noting his many good points, toughness, intelligence, humour. He spoke fluent English, so there would be no need for any but himself, John Ross and CSM Strachan to know his real nationality. Obviously the fewer the better, in case he got put in the bag. Newman seemed to have been everywhere and to have done everything. He appeared to have lived in Paris (he spoke about his mother there) as well as in Berlin, Vienna, Budapest, New York, and London. His background was international, and Frost felt very uneasy about him. 'The Germans then seemed invincible. Their armies knew no halting, and in spite of their recent reverses, or apparent reverses, in the snow in front of Moscow, they were truly formidable. . . . So many things could go wrong with our little party, and we had been taught to fear the enemy's Intelligence. With all the talk in England then and previously about the Fifth Column, I could not help thinking that the enemy probably knew all about us, and what we were training for. There was a distinctly eerie feel to having a Hun on the strength.'[3]

The German's arrival brought the strength up to one hundred and twenty.

Next day 'C' Company with its Royal Engineers, its Flight-Sergeant, and its German interpreter entrained for Scotland.

### NOTES TO CHAPTER 3

1. Cox's written account.
2. Ross, conversations and correspondence.
3. Frost's written account.

# 4
# The Invisible Walls

*1936*

'One of the loveliest places on earth', was Rowe's description of Orfordness, where the first research station was set up. 'At Aldeburgh, in Suffolk the River Ore flows to within a few yards of the sea and then, fortunately, turns south and flows for eleven miles, leaving an isthmus of singular beauty. It was known as "The Island", and those who first worked there on radar were "The Islanders". They had a fascinating job to do, among pink thrift and yellow shingle and the cries of terns.'[1] It was the best time of all. Watson-Watt, a volcano of assurance, led a small group. The preliminary masts were seventy feet high. Experiment led to improvement on improvement until in March, 1936 their new 240-foot tower enabled them to locate an aircraft at a range of seventy-five miles; thanks to Watson-Watt's new discoveries, the cathode-ray tube also informed them of the aircraft's bearing and its altitude. But the high mast was unpopular with a nearby RAF establishment, and the experimental station now needed a more elaborate setting.

Bawdsey Manor, Sir Cuthbert Quilter's place twenty miles to the southward and also on the coast, attracted Wimperis and Rowe because it had extensive stabling and outbuildings in a wooded area of 250 acres. Most of the staff would be able to sleep in Felixstowe, across the River Deben, and get to work by ferry (nothing like a ferry for Security). The pair, sniffing round the boundary fence one afternoon, picked up a rumour that Sir Cuthbert might sell. They heard it at the ginger-beer stall by the front gates . . .[2] The Treasury bought Bawdsey for the nation, and the scientists and their mechanics and tradesmen moved in.

Rowe had a sense of place. He regretted the purity of Orfordness, but he relished Bawdsey's seclusion and style. He liked the peaches, the bougainvillaea, the lawns, the sandy beach under the cliffs, and the motto above the front door which read, *Plutôt Mourir que Changer*.

That spring Bawdsey's lattice steel masts soared above the dark Lebanon cedars and Mediterranean pines and it became operational; for it had been decided to make it as well as the kernel of applied research on Radar the first early warning station of the series now planned to encircle Britain—the Chain Home. Back in 1935 the Treasury had authorised the building of five Radar stations between Bawdsey and South Foreland, to cover the air approaches to London. Even this start was an enormous undertaking for a secret project; the work had to be channelled through public firms, contractors, steel factories, the radio and power industries, the Post Office.

Radar in England now felt its way through political shoals that might have wrecked it, for all its progress and its promise. There was opposition to it in the corridors of power. Professor Lindemann, who had independently urged some means of early warning, disliked his brilliant contemporary Tizard, was initially lukewarm about the Radar programme because Tizard was at the heart of it, and had good reason to suspect Radar's vulnerability.

The Tizard Committee brought Radar to life in England. But soon Ramsay Macdonald, the first Labour Prime Minister, formed another committee, a Sub-Committee of the Committee of Imperial Defence (CID). Its chairman was Sir Philip Cunliffe Lister (shortly to be Lord Swinton), Secretary of State for Air. On June 7 Macdonald departed from 10 Downing Street and Stanley Baldwin moved in. That day Winston Churchill flayed the CID in the House of Commons for its 'lethargy', and also demanded why Lindemann had not been invited to sit on the Tizard Committee. Even out of office Churchill was a force. A month later his scientific adviser, though unwanted, was invited to join the Tizard Committee and Churchill himself attended his first meeting as a member of the politically more powerful CID. Lord Swinton was a friend of Radar, and the argument at the meeting was that Radar must have priority even over fighter aircraft. Wimperis (another member of the Swinton Committee) saw that Churchill sat back listening, a half-smile on his mischievous features, his hand scribbling on the pad before him. And when he slid his scribble across the mahogany Wimperis saw that he had written an excerpt from J. M. Neale's translation from the Greek of a hymn written by Stephen the Sabaite between AD 725 and 794. It is No. 254 in *Hymns Ancient & Modern*. Churchill had taken the first two lines of the first verse:

*Art thou weary, art thou languid,*
*Art thou sore distrest?*

and added to them a version of the seventh verse:

*Seeking, finding, following, keeping,*
*Is he sure to bless?*
*Angels, martyrs, prophets, virgins,*
*Answer. . . .*

but Churchill had altered the bell-like and triumphant final '*Yes!*' of the hymn, to his own, dubious, 'M'yes.' It was an excellent summing up of his own (and Lindemann's) doubts about Radar.

Through the summer of 1936 he and Lindemann were regarded by many as disruptors of the Radar effort. On June 10, Churchill, in a Lindemann-inspired complaint to the Committee of Imperial Defence attacked the Tizard Committee (of which Lindemann was now a member). Two days later Churchill and Lindemann supported Watson-Watt in his attempt to get new management. The inventor complained that progress at Bawdsey was snail-like. Rather, he said, than trying to flog a dead horse (the Air Ministry) into a faster pace, the development of Radar should be entrusted to a new and specially formed organisation. The Ministry was not overjoyed; it was officially felt that progress at Bawdsey had been sensational, rather than slow.

At the mid-July meeting of the Tizard Committee Professor Lindemann rejected the interim report, and Professors Hill and Blackett, deciding that they could no longer work with him, sent their resignations to Lord Swinton. On September 3, Swinton informed Professor Lindemann that the Tizard Committee was dissolved; but the following month he reconstituted it, with Professor V. E. Appleton, uncrowned king of the ionosphere, in Lindemann's place.

Churchill, well-informed as he was, both in and out of office, and aware, as he was, of England's peril, was not hostile to Radar, though he often felt that it was monopolising the country's resources at the expense of conventional rearmament. In scientific matters he trusted Lindemann, who certainly did not dissuade him from acting as a catalyst, from seeking to madden the Tizard Committee and other bodies into 'that period when men cease to be satisfied with present performance and strive to surpass the possible'. And the wrangles suggested above need not influence one against any of the participants including Lindemann, who was often to put England in his debt. Those of us who believe that England's scientists saved her in the Second World War must credit the Prof with influencing the Prime Minister in favour of science and of operational research.[3]

During the pre-war period of quarrels in and out of committee much was achieved. It was, for example, in the name of the Tizard Committee that Lindemann drew attention to the work of Dr Reginald Victor Jones. Jones at Oxford (he was Skynner Senior Student in Astronomy at Balliol) worked in Lindemann's laboratory, the Clarendon. When he was only twenty-three he was approached by an inventor connected with the United States Navy who wished to interest the British Government in an infra-red detection method for aircraft. Jones progressed with this so well that, probably for the first time, one aircraft was recognised from another by infra-red means. He was given the impression, although Tizard invited him to continue his infra-red research at Imperial College, that the Committee wished to close down the work at Oxford because it was being done in Lindemann's laboratory. The Committee made sure however that Jones, even while he continued to work at Oxford, was made a member of the Air Ministry Staff.

* * *

Tizard's unusual value lay in his talent for applying science to service practices. He thought as a scientist and also as the pilot he had been. (Lindemann too had been an outstanding pilot.) It was not enough, Tizard felt, for the Royal Air Force to have warning when hostile aircraft approached; its reactions to such attacks must be trained reactions. However successful the new British fighter designs (they were having their troubles) and however rapid their production (it remained disappointingly low), the system of airborne patrols must be changed. The new, much faster, more heavily armed fighters must sit on their home fields ready to spring at the intruders. They must be guided to within sight of the enemy by controllers who, watching the enemy and their own aircraft with Radar, could bring the two together. This seems mere common sense today. In the summer of 1936, when Tizard was allowed to carry out some experiments at Biggin Hill, it was a pipe dream. (And time was running out.)

Biggin Hill was the central airfield of the London defence perimeter. Wing-Commander E. O. Grenfell acted as host and worked with

Tizard's small group. Tizard had asked Rowe to pick him a young scientific assistant. The choice was Dr B. G. Dickins, then working at Farnborough, and they were joined by Squadron-Leader R. L. Ragg, another from Farnborough, and Flight-Lieutenant W. P. G. Pretty, a signals officer. They had three Hawker Harts to represent enemy bombers. Their fighters, three Gloster Gauntlet biplanes, waited at Biggin until ordered off and then were guided by radio to the interception. The trials showed that *if* a fifteen-minute warning was given, and *if* the controller was good and had first-rate R/T communication with the pilot, and *if* the fighter had a performance equal to or superior to that of the bomber, the system worked.

After a fortnight of interceptions Dickins suggested to Tizard that it would be better to have R/T contacts at one-minute, rather than five-minute, intervals. Tizard then got the Air Ministry's authority to tell his helpers that Radar existed and that Bawdsey was making contact at a hundred miles and more.

When straightforward interceptions were becoming easy the 'bombers' were ordered to jink, to change course. The defender's original course then had to be altered in midflight, and the controller had to work fast. Here Tizard came in once more. He produced a Principle of Equal Angles: at any stage of the interception a line could be drawn between fighter and bomber, and this was the base of an isosceles triangle; the prolongation of the bomber's course was the second side of the triangle, and the course to be given the fighter made the third side. This formula became known in the RAF as 'The Tizzy Angle'. The exercises continued until one day the controller, Grenfell, saw that a mistake had been made in the fighter courses. There was not time to work out a new Tizzy Angle. He therefore corrected by eye, and made the kill. Tizard deduced from this that an experienced fighter controller would do much of his work by instinct. The scope of the experiments was then widened, with more aircraft and more controllers. It was found that the best controllers could manage four interceptions simultaneously. A new language was evolved where *scramble* meant take-off, *vector* meant compass course, *angels* meant height, and *pancake* meant landing. Almost a year's work at Biggin Hill changed the form of defensive air war. Now others would step forward to refine and expand the new law of the air that Tizard and his Biggin Hill group had made.[4]

* * *

'Pip-Squeak' was an Air Force gadget dating back to 1932. Dowding called in Watson-Watt and the Bawdsey scientists to develop it and to help improve the high-frequency direction-finding that, following Tizard's experiments, was seen to be needed. Dowding asked for three High-frequency direction-finding (HF/DF) stations in each Fighter Command sector. The fighter pilot's 'Where am I?' of the (good) old days was to be taken away. Ground must know exactly where he was and Ground would direct him either on to the enemy or home. It was, initially, an unpopular notion with the pilots. 'Pip-Squeak' automatically switched on the HF transmitter in the fighter for fourteen seconds in each minute of flight. The pilot had an over-ride control on the device should he need his radio during the squeaking seconds. A clock with its face in four coloured sectors was duplicated in the control room and the DFing stations. The clock told them which aircraft was pip-squeaking and the position of each fighter was fixed each minute by plots from the three stations.

Simpler, and perhaps more romantic if only because it called for human endurance, was another most important part of air defence, the Observer Corps. It was a peculiarly British organisation, a mass of superior volunteers hung on a skeleton of professionals. The Corps had operated in 1918, lapsed after that war, and began to form again in 1924. It was first run by the police, and fortunately when the police handed over to the Air Force it was kept as a civilian body. In Germany the *Flugmeldedienst*, a less effective, uniformed service, functioned in the later years of the war; in Britain the harder the civilian Corps worked the better it became. The map of the Observer Corps is an impressive sight, covered, apart from gaps in Cornwall, West Wales, and the West of Scotland, by the five-mile-radius circles of the section posts. There were some thirty posts to each centre, and thirty-two centres, divided into five zones, were connected to Dowding's Fighter Command H.Q. at Bentley Priory, Stanmore, Middlesex.

The Observer Corps was at full strength in 1938 when the Munich crisis came and Hitler, fortunately for Britain, if not for Czechoslovakia, decided to buy more time with lies. The Corps, like the Chain Home, was called out for a full-scale stand-by. Both systems were ready for action, but the Corps was the readier. At that time the General Post Office had all but finished the immense installation of direct lines from the posts to the command centres of the Corps. The GPO had been laying land lines for the RAF since the inception of Radar, and in the year following Munich it doubled the mileage of all those lines. In July, 1938 the Post Office began to set up its own child, The Defence Teleprinter Network (DTN), a communications system that carried most of the important routine business of the Battle of Britain, indents, damage assessments, supplies, casualties.

By the outbreak of war the British had completed an early-warning defence. The twenty final Chain Home stations had 350-foot lattice-steel towers for the transmitters and 240-foot wooden towers for the receivers. They could range, fix, and read the altitude of aircraft 120 miles distant, and they could normally distinguish their own from hostile aircraft. Advance stations had 90-foot steel towers and mobile equipment. Intermediate stations had 240-foot towers and mobile or experimental equipment. In addition, before the German air attack began, thirty Chain Home Low (CHL) stations were operational. These were coastal defence Radar sets with revolving aerials. They had been designed by Bawdsey and developed for the Royal Navy, to locate and range hostile ships and submarines. The Air Force, however, saw the set as a possible answer to low-flying marauders. The Radar floodlight or twilight transmitted by the huge masts of the Chain Home did not descend right down to the surface of the sea, and it was possible to fly under it. It was felt that Chain Home Low would close that gap. The installations were rushed through under the leadership of a Cambridge don, Professor John Cockcroft. In addition the Observer Corps gave 'Low Raid Urgent' priority to low-flying aircraft.

So, if the British had failed in many respects to ready themselves for the fight against Nazi Germany that had become inevitable in the 1930s, they had done wonders with their early-warning system; they had been as thorough and as painstaking as they had been inventive. Its like existed in no other country. Invisible walls had been built round the United Kingdom, walls twelve miles high and one hundred and twenty miles thick. H. G. Wells himself could never have imagined such defences. . . . And the gallant young men in their flying machines were there to hold the walls.

Oddly enough, although the United Kingdom's efforts and expenditures in making these installations were proportionally as great as Egypt's when the Pyramids were built, the Germans still knew nothing about British Radar. Nothing at all.

NOTES TO CHAPTER 4

1. Rowe, *One Story of Radar*, p. 13.
2. Rowe, *op. cit.*, p. 20.
3. See Birkenhead, *The Prof in Two Worlds*, Harrod, *The Prof*, and Clark's *Tizard*, and *The Rise of the Boffins*, pp. 43–9.
4. Wallace, *RAF Biggin Hill*, pp. 92–5; Clark, *The Rise of the Boffins*, pp. 49–54; etc.

# 5

# Loch Fyne Interlude

*January, 1942*

At the miraculously beautiful little town of Inveraray on Loch Fyne, in the coastal fringe of Argyll, 'C' Company and its appendages were swallowed up by the Royal Navy. They were accommodated aboard HMS *Prins Albert.*

*Prins Albert* had been built for the Belgian Government by John Cockerill's at Hoboken in 1937. At the outset of war Harland and Wolff had done a conversion, and the ship was commissioned for special service in September, 1941. Her 15,000 horse-power diesels gave her a maximum speed of 22 knots on a length of 370 feet, a beam of 46 feet, and a draught of 13 feet. She could carry eight Assault Landing Craft. Her armament consisted of two twelve-pounders, two two-pounders, and six Oerlikons and she was manned by thirty-five officers and 161 ratings. 'C' Company fitted into her guest accommodation most comfortably. She had quarters for thirty-nine Army officers and three hundred other ranks. After the mud and huts and duck-boards of Tilshead the ship seemed a floating hotel, spotless, hospitable, and with excellent food and drink. There were *eggs in the shell*, something that they (with their favoured rations) had not seen for months.

'Even the subalterns had enough to eat,' Frost says. 'And there was no shortage of the civilised things in life. We were lucky with the weather, and all of us greatly enjoyed splashing about in the landing craft, even if it meant long hours and frequent wettings in the icy waters of Loch Fyne.'[1]

Assault Landing Craft (LCAs) were of Thornycroft design and were built in yacht yards. They were 41 foot long, with 10- foot beam, and with a draught of only 2 foot 6 ins. Two Ford V8 engines gave, on flat water, a supposed 10 knots, and each carried a crew of four, and a maximum of thirty-five fully-armed marines or soldiers.

Although Frost had not yet been authorised to say a word to any of his officers or men about the purpose of their training, and the cover story that they were to do an exercise in order to convince the Prime Minister was still maintained, it is doubtful if any of them failed to understand that something more urgent lay ahead. So the wooden-hulled boats with the quirks of their petrol engines endeared themselves to the soldiers. They represented perhaps the difference between glory dead and glory alive.

Operationally, though, Inveraray was one prolonged headache. They were disconcerted to find that night embarkations off the rock-strewn beaches were extremely difficult and, if the weather piped up, hazardous. There was always that phenomenon, the tide. Sometimes it was making and sometimes ebbing; but always it seemed to be wrong for what they wanted to do. Then whether it was coming in or going out, it was going sideways as well. Also (stone the crows!) the naval crews were fallible. They had the greatest difficulty even on those occasions when radio contact was well established in finding the dark figures on the darker shores. They never appeared to be able to distinguish torch signals, and they often missed coloured Very lights. All in all, there had not been a single successful night embarkation, and Frost was getting increasingly worried by the failures when he was asked in the Ward Room if he and his blokes could make themselves scarce the next morning as Admiral Lord Mountbatten was coming to inspect the ship.

Frost and his men were accordingly busy with an exercise ashore when they heard the ship's whistle and saw the landing craft leaving her flanks. At the same time messages came through on Company Headquarters' 38 set that the Admiral urgently required their presence.

Mountbatten addressed all ranks of both services (plus Flight-Sergeant Cox). Until that moment the Navy had not known that its 'Pongos' were parachutists, or that what they were trying to do there was to be repeated in action in which *Prins Albert* would take part. The Admiral made it clear that a military exploit of small scale but of great significance depended upon absolute co-operation between the three services.

At Tilshead 'C' Company officers had been visited by a good many outsiders as well as by their own sort from the First Airborne. Politicians and scientists, soldiers and airmen, had made much of them. Mountbatten struck them as being business-like, matter-of-fact, and reserved. He saw Frost alone, after the meeting. Both were young men, the Admiral unusually so for the responsibilities he carried. He asked Frost if he had any doubts or worries. Only two, was the answer; firstly, that the problem of rescuing 'C' Company, the Bruneval raid accomplished, was certain to be a much stiffer one than he had supposed; and secondly, he was unhappy about 'Newman'.

Private Newman was at once sent for, and Lord Mountbatten subjected him to what Frost admiringly considered to be 'a tremendous barrage in absolutely fluent German. Lord Mountbatten stood him up in front of the desk and shouted at him. He answered quietly and smoothly. They shook hands, Newman was dismissed, and the Admiral turned to me. "Take him along," he said. "You won't regret it for he's bound to be very useful. I judge him to be brave and intelligent. After all, he risks far more than you do, and of course he would never have been attached to you if he hadn't passed Security on every count." ' As for 'Newman', he remembers Lord Louis's interrogation on the ship as a pleasant and gentlemanly affair. Hewas asked about his life before the war and his reasons for wanting action. He thought the Admiral utterly charming.[2]

Next morning *Prins Albert* steamed down Loch Fyne and the parachutists caught a train from Gourock for Salisbury. They were bound for another dose of Tilshead Camp, and hoped the Glider Pilots' Regiment would be glad to see them.

As on their previous arrival at Tilshead, they were to meet General Browning the following morning, but this time they were to call on him, by parachute. Frost subsequently confided his disgust to his diary:

'It's a bombing squadron with a fine operational record. But it has never dropped live parachutists. There was disorder and confusion and a mighty waste of time on the aerodrome—a shambles in fact. It was not until late in the day that we tumbled out on to a piece of brick-hard ground right in front of Syrancote House. By a miracle there was, thank God! no grief, not even a twisted ankle. The General said he was satisfied. More than I was. . . . Sergeant Grieve made the rest of us look slow, he's taught his Seaforths to deplane so amazingly quickly—yet another thing to get right. . . .'[3]

NOTES TO CHAPTER 5

1. Frost's written account, and conversations.
2. Conversations with 'Newman' and Frost's written account.
3. Frost's written account.

# 6
# A Confident Major Schmid

*1939/40*

In the field of Intelligence the British had long been skilled at calling an apple a turnip, a turnip a diamond, or (as in this case) a diamond a turnip. It was clearly impossible to hide twenty Chain Home stations with multiple 350-foot steel towers; so they were passed off as radio stations. What kind of radio? A story was disseminated, and was credited in Germany and in England, that the Royal Air Force was being strategically wrecked by a too-tight system of radio control from the ground. The Luftwaffe lost no opportunity of studying the RAF. The *esprit de corps* of the British service and its smartness, the practical neatness of its airfields, the officer training at Cranwell, the thoroughness of its ground engineers—all these were admired in Germany, but . . . the Luftwaffe had had the new kinds of aircraft, heavy metal-winged bullets, before any other nation. It knew it was the best Air Force in the world. (It was still telling itself that it was the best when it had perhaps ceased to be so.) Its fighter pilots had enjoyed a taste of active service in the Spanish Civil War that had taught them much about tactics with the new type of aircraft. They noted with scorn that the RAF (which had not had the advantage of a Spanish try-out) was still clinging to the old-fashioned close-knit fighter formations that looked so attractive at air displays.

Throughout 1937 and 1938 England was overflown by unarmed Heinkel IIIs. Officially these aircraft were carrying out timetable and weather flights on behalf of Lufthansa. In reality they were doing photo-reconnaissance for the Target Information Department of the German Air Ministry in Leipzigerstrasse, Berlin. German bombers as a result were well documented at briefings-to-come for the attack on England. Chain Home stations were photographed and described. It was noted that the ancillary buildings were tucked in under the masts and protected by sandbags and blast walls, and that the masts themselves were enormously high. They would make interesting strafe targets. The Stukas would enjoy them. But priority was given to pictures of naval dockyards and of the bigger commercial ports, London, Bristol, Liverpool, Glasgow. Second on the list (showing the current German undervaluation of the Royal Air Force) were RAF targets, all the airfields south of London and west to Plymouth and Bristol, and the many big ones made or making for the British bomber bases in East Anglia. A special section of the photographic cover—and this was particularly well presented and annotated—dealt with the shadow factories producing aircraft and aircraft components and weapons.

General Wolfgang Martini, commanding Luftwaffe Signals, could not reconcile the masts with Dezimeter Telegraphie. Even so, he knew as a conscientious officer that something should be done about them. Accordingly he got agreement from the Air Staff for the employment of the *Graf Zeppelin*, LS 127, once Germany's pride but now rather old hat, as a flying radio laboratory. The *Graf*'s gondola was packed with high-frequency receivers, and an aerial array was fixed below.[1]

General Martini was aboard when at the end of May, 1939, *Graf Zeppelin* unhooked from a Friedrichshaven mooring and crossed the North Sea, making landfall at the mouth of the Deben. At Bawdsey (now withdrawn from the Chain Home, and merely a research Radar station) and Canewdon (Chain Home) the biggest imaginable blip appeared on the Radar screens and moved majestically across them. The English tracking system, delighted to have such a jumbo practice run at the German taxpayer's expense (for a change), followed the *Graf* mile by mile as it thrummed its way north up the coast. The weather was thick, and off the Humber (as was thought) the airship reported its position to Frankfurt. At Fighter Command in Bentley Priory this caused amusement, as the German navigator's reckoning was nine miles east of his true position; he was actually flying over England. The flight continued up the coast to the North of Scotland, then turned for home. It was a complete failure. All that Martini and his crew received for their pains was loud static. They thought the trouble might be caused by installation faults or by reflections from the skin of the airship.

A second run was made on August 2, 1939, in thick weather and high winds. The *Graf* cruised from Suffolk to the Orkneys, and aroused a good deal of uneasiness, for she was sighted intermittently by coastguards and Trinity House men, yet this time there was no trace of her on the Radar screens. Were the Germans, then, operating some form of jamming? We shall see later that when he wanted to be, General Martini was a jamming artist, but it seems unlikely that he had tried such a thing to shield a simple scouting flight; for one thing, he did not yet know the Chain Home frequencies. The second flight was also abortive as a radio search. This made the Germans wonder if the British used jamming to hide whatever those tall masts were transmitting.

One more flight was made by the *Graf*, but again no British transmissions were picked up.

* * *

Abwehr (meaning defence) was originally the name of the counter-espionage organisation of the German Army. From 1921 the Abwehr, which had barely survived Germany's troubles after the First World War, was steadily and carefully expanded, and its duties came to include offensive as well as defensive intelligence, in Germany and abroad. The service was deliberately decentralised. Its officers were chosen with care, and were of the educated, upper class, officer type. Its biggest station outside Berlin was in the great seaport of Hamburg with its Anglo-American and international connections. In the Europa-Saal suburb of Hamburg the secret wireless stations of the service were built, the transmitting station some distance from the receiving one. Highly trained operators manned the stations on twenty-four hour rotas. For communications and for detecting, filing, and decoding enemy transmissions, the wireless work of the Abwehr was of a high standard. This was not surprising as the services chief, Wilhelm Canaris, was of course an Admiral. (He had ended his long career in the Navy with the rank of Captain.) The son of an industrialist, with no Greek but some Italian blood in him, Canaris was what the English call a gentleman. He was a reflective man, in many ways a moralist, fiercely anti-Communist, a leader who gave his men much scope, and who expected and obtained from them unlimited devotion.

In the fields of sabotage and 'fifth column' operations, particularly at the beginning of the war, the successes of the Abwehr's 'Brandenburg Commandos' were phenomenal. In the more classical sides of its work, intelligence and espionage, it would appear to have succeeded in France, the Low Countries, Spain (Franco and Canaris were friends), Italy, Scandinavia except for Sweden, all the Central European countries, the Balkans, and the Middle East wherever British cover was weak. It would appear to have failed in the United Kingdom and the United States.

Those failures, or partial failures, are often blamed on the hostile atmosphere in which the Abwehr had to work. For as soon as the Nazis' own security service under Himmler, the Sicherheitsdienst (or SD) grew strong it began to steal power from the Abwehr which finally, in 1944, it was to break in pieces by sending Canaris to a concentration camp where he was murdered. As one of the Abwehr's former officers, Paul Leverkuehn, has said: 'This political organisation (the SD) had been one of the major factors in limiting the efficiency of the German Secret Service. . . . Once Canaris had gone, the Abwehr soon began to disintegrate . . . while the Nazi SD, using methods which to Canaris were anathema, tried with only limited success to fulfil the functions of an Intelligence Service.'[2]

German espionage in the United Kingdom was severely restricted (as compared with Europe), even in the 1930s. A strong mistrust of Germany existed, following Britain's terrible losses in the First World War, and this increased, for all the country's apparent torpor, as Hitler's power and his evident barbarity grew. The German governess or tutor had gone out of fashion, the German Hospital and centres of German trade in London were closely watched, and counter-espionage both in Britain and in North America was much hotter than in other countries. Then the Ambassador in London, Ribbentrop, was hostile to the Abwehr and insisted on running his own private Intelligence Service, which was penetrated by British counter espionage. After the outbreak of war, and particularly after Dunkirk, the United Kingdom bristled with arms and its naturally suspicious and xenophobic inhabitants became more watchful than ever. The parachuting method of infiltrating agents, which the British used so successfully in Occupied Europe, was made nearly impossible for the Germans by the round-the-clock and all-embracing watch of the British Observer Corps.

Abwehr cells in the United Kingdom there must have been, and some of them must have been able to communicate with the Europa-Saal in Hamburg. But such cells cannot have got near the roots of British power or British secrets. Of course when a strong Intelligence complex finds itself on the losing side in a war it simply covers up and waits patiently until things change, as they always do. Possibly a few Germans know today what Abwehr (or even SD) cells survived the war in

Britain. Of this we can be sure, no Intelligence source informed the German High Command of the remarkable extent to which the British had transformed their air defences. The High Command suspected before the outset of war that the British possessed some crude form of Dezimeter Telegraphie. Probably this suspicion arose from some report from Abwehr One TLw (Luftwaffentechnik) Section; but one of the Abwehr's increasing worries was that as political hostility to it grew, its reports tended to be ignored or disclaimed by politico-military figures like Göring.

Göring indeed now had his own intelligence centre, which had been formed by the amalgamation of the Target Information Department and the Foreign Air Forces Department, formerly two separate sections of the Air Ministry. Abteilung Five, the new amalgam, was commanded by Major Josef ('Beppo')[3] Schmid, a member of the Party and a protégé of the Air Chief of Staff, General Jeschonnek. Major Schmid apparently knew, as did many another German bureaucrat and fighting man, that Hitler and Göring, though formidable, were avid for good news and readily angered by bad. For a long time Major Schmid, since obviously the Luftwaffe was vastly superior to any other Air Force, could conscientiously hand out only good news and forecast that the omens, too, were favourable. (The information issued by that department often had a Wagnerian smack to it.)

When Schmid got down to work the Germans already knew a great deal about the French Armée de l'Air, indeed they knew more about French military and political weaknesses than did France's main ally, Britain. General Martini knew for example that DEM (Détection Electro-Magnétique), the French early-warning device, was quite useless. What Martini did not know was that the British suspected DEM and offered to share their own Radar system with their French ally. On May 23, 1939, a French military mission to London was shown the subterranean filter room at Bentley Priory. The British Radar machine had then been tested and retested, changed and perfected. To say that the French officers were flabbergasted would be an understatement. Early in June six French radio specialists, two of them from each service, came to England and went through the full Radar course including a study of manufacturing, installation, and maintenance. A scheme was promulgated in France to erect Chain Home stations, but like so many other schemes at that time (and earlier) nothing had come of it before the coming of the Germans. If Major Schmid had not been informed of that, he was able to read in up-to-date Intelligence reports of the irreparable damage done to France's strength by the Blum Socialist Government, when Pierre Cot nationalised and all but wrecked a strong aircraft industry. He read too of the ineffectuality of an elderly French General Staff that commanded respect everywhere in the world bar Germany and Russia. Russia, like Germany, was properly informed on all aspects of life, commerce, and armament in France, and should Russia be on Germany's side *at the beginning* her agents could and would play a big part in the internal affairs of France.

Schmid was required to assess the air strengths of Poland, Russia, and Great Britain. The first two were comparatively easy. And he managed to get out a long report, *Studie Blau*, about the Royal Air Force and British air defence and bomber strength. His sources were not numerous. Generals Milch and Udet, for example, had made a successful official visit to England in 1937, when they had inspected RAF units, and some of the air industry's factories and shadow factories. Both Generals, but especially Milch, who was Göring's second-in-command, had been impressed. Milch had gone so far as to say that in time the Luftwaffe might find the RAF a hard nut to crack. (Göring was reported to be furious.) Otherwise Schmid's sources were standard ones, air reconnaissance, the London air attaché's office, and the excellent English technical and semi-technical periodicals connected with aeronautics and the air industry. Throughout the Battle of Britain *Studie Blau* held its place as a German text book. *There was no mention in it of the new methods of warning, communication, and command that had been developed by the British.*

* * *

'Born of the spirit of the German airmen of the First World War,' Göring began his Order of the Day before the advance on Poland, 'inspired by faith in our Führer and Commander-in-Chief, the Luftwaffe today is ready to carry out every command of our Führer with the speed of lightning and undreamed-of power.' And Major Beppo's victory calculations were confirmed. The Luftwaffe smashed the Polish Air Force, four hundred aircraft, in two days, many of the Polish machines being caught by Stukas on the ground in the first twenty-four hours.

An unusually hard winter then precluded further German attacks.

The Polish campaign had been systematically used and analysed as an exercise as well as a conquest. The Germans now believed that they understood the tactics of a new form of war. One key to it, as in the Napoleonic and other wars, was good weather; but now the reason was different, since the main striking force was their Air Force. The spring weather forecasts were entirely favourable. On April 9, 1940, German armour violated Danish neutrality at five in the morning, and by that evening Denmark was conquered, its Air Force strangled by sudden attack. Norway was a degree more complicated. The Luftwaffe was surprised to lose in taking Norway fifty-four bombers and thirty-five Ju 52 three-engined transports.

In May it was to be the turn of Belgium, Holland, and France. And it was a wonderful month of May. The sun that shone was a German sun. For France it pitilessly illuminated her unsuspected military decay. The German plan for attack was based on achieving complete air superiority in two days. This was done by the Second and Third Luftflotten, and both air groups then turned to army co-operation, the role for which they had been designed. The Eighth Fliegerkorps' four hundred Stuka dive-bombers performed as mobile artillery for the German armoured thrust aimed at the Channel coast.

For the Royal Air Force the month from the beginning of the German attack to the end of the Dunkirk evacuation was the most painful of the whole war. Committed by the Supreme Commander, General Gamelin, to a hurried advance into Belgium, the RAF, like the British Expeditionary Force, was caught in the worst of the chaos. The squadrons had no early-warning system. They were out-classed and outnumbered by a Luftwaffe acting 'with the speed of lightning and undreamed-of power'. Only six squadrons of Hawker Hurricanes were operating initially in France, and the British light bombers consisted of the hopelessly vulnerable Fairey Battle, and the Bristol Blenheim, adapted from a civilian aircraft, patriotically sponsored by Lord Rothermere, and quite unsuited to that kind of fighting.

Göring, fortunately for England, took a hand at Dunkirk. Hitler was nervous about using his armour there, and gladly accepted Göring's assurances that the British evacuation could be extinguished from the air. Göring had five hundred and fifty fighters in the area, most of them the formidable Messerschmitt 109s, and he could deploy the bombers of the First, Second, Fourth, and Eighth Fliegerkorps. Even incommoded by the clouds of smoke that often shrouded the area, and the suddenly fierce ground fire of concentrated British troops, and the hostile reaction of such ships as were armed, the Luftwaffe should have contained the evacuation. That they did not was a victory for the RAF, but neither the British nor the Germans regarded it as such. Dowding used his Spitfires for the first time over Dunkirk. His fighters were operating outside their HF/DF and Radar system. They had to mount old-fashioned defensive patrols, and they were still flying the tight pre-war formations. British fighter losses were heavier than German. But the evacuation was completed.

* * *

In her greatest days, the conquerors claimed, England had never won a victory to compare with Germany's 1940 triumph. Paris had fallen. Hitler was there, in Paris. France, Belgium, and Holland were outstandingly rich and attractive countries. Nobody gave exceptionally high priority or urgency to the planning of Operation Sealion, the invasion of England. But the Luftwaffe flowed in like quicksilver to the Channel airfields of France.[4]

The Second, Third, and Fifth Luftflotten were to handle England. Field Marshal Albert Kesselring, commanding Luftflotte Two, had his headquarters in Brussels, with advanced headquarters at Cap Griz Nez from which the four immense towers of Chain Home Dover were usually in full view. Luftflotte Three, commanded by Field-Marshal Hugo Sperrle, hadheadquarters in Paris, and advanced ones at Deauville. The outline plan was that those two Luftflotten, victorious in the battle just finished, should divide Southern England, Sperrle taking the West Country, Kesselring the South-East, while General Hans-Jürgen Stumpf, commanding Luftflotte Five, would fly across the North Sea to attack Midland targets and Southern Scotland. But even in the Luftwaffe June was a month of détente, of triumph enjoyed. There was no realisation that the hard work was only about to begin.

In May the Germans had captured at Boulogne a British mobile Radar set. Analysis of enemy aviation equipment was the responsibility of General Udet's Air Production Ministry, but General Martini's department was asked to give an opinion. It was decided that the British set was crude and ill-made, and that it worked on an impractical wavelength. In the same period Udet's department examined complete specimens of the Hurricane and the Spitfire and declared them to be of poor quality and technically inferior to *both* Messerschmitt fighters. They appeared to base this assessment on the lack of cannon armament (at that time) in the British machines, the lack of armour behind the pilot, and the fact that neither had fuel injection. It was true that at high altitudes fuel injection gave the 109 an advantage, since it could go straight into a dive without risk of cutting out; but no consideration was given to the high rate of fire and the reliability of the eight Brownings in the wings of each British fighter, or to the handling qualities and the ability of both Hurricane and Spitfire to take punishment and keep fighting. Udet's findings were passed on to Major Schmid, and were publicised. Radar, the Hurricane, the Spitfire, all dismissed as inferior! It was this kind of wrong-thinking in Intelligence that was to distract the German pilots during the Battle of Britain. As for the twin-engined Messerschmitt 110, Göring's 'destroyer', far from being superior to Hurricanes or Spitfires, it was all but helpless before them, and its failure, since it had a longer range than the redoubtable 109, was a severe blow to the Luftwaffe. Another failure of German Intelligence was at the Air Ministry in Paris, where the vital lead into British Radar secrets was totally missed. The French Air Ministry, contrary to their security promises to the British, had actually put out tenders with commercial contractors for the erection of Chain-Home-type stations on the Biscay coast.

By the end of June Göring had ordered the Second Fliegerkorps commanded by General Lörzer and based on the Pas de Calais, and the Eighth, commanded by General Richthofen at Le Havre to sweep the English Channel and obtain and keep air supremacy. Lörzer delegated the post of Kanalkampfführer to General Johannes Fink. He put under Fink two Stuka Gruppen and two Me 109 fighter wings JG 26 and JG 53. These last were led by aces, Major Adolf Galland and Major Werner Mölders, idols of the German people: hot opposition indeed for the RAF pilots in their 'inferior' machines.

General Martini had come to the Channel coast with his Signals sections and some German Radar had moved in. There were two *Freya* stations in Northern France, and a *Seetakt* had been installed near Cap Griz Nez to sharpen the gunnery attacks on British convoys negotiating the Straits of Dover.

Action in war, where two strong combatants are involved, is inclined to be self-stoking, and while Dowding jealously husbanded his fighters in the convoy battles, knowing quite well what the enemy was after, and hating his high-flying Me 109 cover, Fighter Command was kept so busy that the British regarded July 10 as the beginning of the Battle of Britain. Naturally enough, the Germans were over-confident. Major Beppo Schmid's optimism went right through the service. Victory might be bloodier than over other conquered countries, but it was inevitable.

Then a discordant voice was raised. It was the voice of General Martini. Now that there was a great deal of air activity (even if the official Eagle Day had not yet been fixed) Luftwaffe Signals' monitors had had a surprise. They had never imagined anything like it. On the twelve-metre band the ether simply pullulated with signals emanating from the tall masts of the Chain Home—the Radar twilight. And when the German squadrons flew northwards across the water Martini's men heard on the high frequencies the process of RAF fighter squadrons being talked into battle. It was evident, though how it came about was yet a mystery, that British officers sitting somewhere near London were actually 'seeing' the Germans take off from runways lying well back in France. They obviously thought they knew exactly where the German attackers were, and they also knew exactly where their own fighters were, even when they were airborne. The monitoring staffs also heard the pilots answering the controllers and talking to their companions. And often the British leader would cry into his microphone, 'Tally ho! . . . Heinkels with 109s . . . Tally ho!'

Martini and his staff knew they were hearing something important; that the British had developed an original form of aerial defence. But the High Command refused to be ruffled. Worry made cowards. Boldness brought victories. Once more Schmid's Abteilung, in an appreciation of the British defence issued to the three attacking Luftflotten and the two Fliegerkorps, told the senior officers what they wanted to hear. 'Because the British fighters are controlled from the ground by R/T their units are tied to their respective ground stations and are therefore restricted in mobility even if, as is likely, the ground stations are sometimes mobile. It follows that the forming of a strong fighter force at crucial points and at crucial times is unlikely. There will be confusion in the defence during mass attacks . . .'

This facile and erroneous interpretation of the Radar twilight, the chatter, the pips and the squeaks was to rebound on Abteilung Five particularly in conjunction with Five's habit of publicising exaggerated claims of British losses.[5] What was taken to be 'restrictive' control was in fact a control that gave the outnumbered British pilots some time to rest on the ground and that could draw them rapidly from a wide area to a narrow front. German pilots, continually hearing and reading that hundreds of Spitfires and Hurricanes had been destroyed, became exasperated to be met on nearly every sortie by more and yet more Spitfires and Hurricanes. How many of the damned things did the British have, anyway?

And Martini's monitors came to dislike very much the all but incomprehensible phrase, 'Tally ho!'

* * *

Göring told his entourage, and Hitler, that all he needed to finish off England with his Eagles was one week of such glorious weather as they had enjoyed over France in May. Any English farmer could have told him that August is one of the most fickle months in (even) the English calendar. In the end he lost patience, and named August 13 as his Eagle Day, the official start of a fight that had been going on for more than a month. The day before, the twelfth, was a dangerous day for England.

Heavy and repeated German attacks across the Channel began at seven-thirty in the morning. At nine o'clock five British Radar stations were bombed, and soon after fifteen Ju 88s skilfully lacerated Ventnor Chain Home station on the Isle of Wight. Every building at Ventnor was damaged and on fire. Dover, Pevensey, and Rye stations suffered badly. The controls, ancillaries, and power lines at Ventnor were in a terrible mess. Every other Radar station was operational by nightfall, but Ventnor could not be resuscitated.

On Eagle Day the Germans, who did not realise how well they had done the day before, flew 1,485 sorties, and things went ill for them. Although Luftwaffe Supreme Command (OKL) claimed seventy Hurricanes and Spitfires and eighteen Blenheims destroyed, the true figures were thirteen RAF aircraft destroyed and forty-five German. Many of the forty-five were Stukas (Ju 87s) and from the Radar viewpoint that was important. Until this battle the Stuka had been a war winner, but what happened in just five minutes on Eagle Day was typical enough. . . . Thirteen Spitfires of 609 Squadron saw a formation of Stukas heading for Middle Wallop aerodrome under a cover of Me 109s. The Messerschmitts were already involved with other British fighters. The Spitfires dived through them, bringing one 109 down and destroyed *nine* Stukas.

August 15 was a field-day for the Chain Home. OKL decided to attack on three fronts. General Stumpf sent bombers across the North Sea. The Chain Home picked up all the North Sea 'hostiles' at full range and squadron after squadron of fighters went up to rend them. Fighter Command on that day flew 974 sorties, lost thirty-four aircraft, and claimed 182 Germans, but actually shot down seventy-five.

There were grim faces at Karinhall, where Göring was upsetting his junior commanders at an irascible conference. That day the Luftwaffe had lost every type of aircraft that was being used against England (Do 17, He 59, He 111, He 115, Ju 87, Ju 88, Me 109, Me 110, and Arado 196). Göring, having called for a frank discussion, chose to blame his fighter pilots for failing to protect their bombers, and particularly the Stukas, which he loved. He was infuriated by their reports that the Me 110 was no match for either British fighter, and that at low altitudes the Hurricane was as good as, and the Spitfire possibly superior to, the Me 109. The 109, the pilots also reminded him, had the shortest range of any fighter, some 420 miles, which hampered it for the kind of close escort work he now wanted them to fly. Tactically too, they said, close escort work put the 109 at a disadvantage with the agile British fighters. Göring nearly, and most unjustly, accused them of cowardice, but changed the subject. Another infuriating aspect of the British campaign: photographs showed that British forward airfields reported 'destroyed' or 'obliterated' were still being used. Photographs also showed, and this was indeed alarming, that German daylight bombing under pressure was not precise and reliable, as it had been in the other campaigns when air supremacy had been instantly won. He pointed this out to the fighter pilots. It was up to them to see that the bombers got a clear run. . . . He turned to the question of losses. Too many Luftwaffe officers had died or had been shot down over England; from that day on there was to be a maximum of one officer in each aircrew. . . . He was asked if he stood by his directive that, in the night-bombing of England the special squadron, Kampfgruppe 100, should be used. Yes, Göring answered, yes, it must be used. It was pointed out that there was a lot of secret equipment in each Heinkel 111 of Kampfgruppe 100. Equipment! There had been too much talk about British equipment. Wars were won by men, not by equipment. From England's point of view, indeed from any point of view, the most important of Göring's decisions at that important conference concerned Radar. 'It is doubtful,' he said, 'if there is any sense in continuing the attacks on D/T stations, since not one of those so far attacked has stopped transmitting.' (Engineers, contractors, electricians, scientists, RAF specialists, and GPO secret squads were toiling on the Isle of Wight, yet Ventnor CH was still off the air.)

Even if Göring did think it a waste of time, the following day Ventnor was again dive bombed and was again severely mauled. It was not until seven days later, when a mobile station began to operate at Bembridge, that the Ventnor gap in the Chain Home system was plugged.

The Second and Third Luftflotten attacked savagely across the Channel on August 18. Poling Chain Home on the serene Sussex Downs behind Littlehampton was lashed by Stukas. Ninety bombs fell inside its perimeter. But no other Radar stations were attacked, and again Stukas were lost, twelve of them inside ten hours. And the hitherto victorious angular two-seater with its spats and its vulture profile was pulled out of the Battle of Britain. Six Stuka Wings had been lined up for the attack, but by the end of that crucial month, August, they had all been withdrawn to be used on other fronts where the opposition was less hot.

So the Battle burned on and Radar played its essential part until the attacker, always under pressure and ill-prepared mentally to settle down and fight coolly against such resistance, made the final error of turning to attack London.

## NOTES TO CHAPTER 6

1. Description of Martini's radio searches with the *Graf* are based on the Prologue of *The Narrow Margin* by Wood & Dempster, pp. 17–20.
2. Leverkuehn, *German Military Intelligence*, p. 195.
3. Wood & Dempster, *op. cit.*, p. 101.
4. At this stage General Erhard Milch, who had probably done more than any other officer to make the Luftwaffe the great and successful service it was, flew his own Dornier to Göring's headquarters in Belgium, to urge that the momentum of German victory must *immediately* be continued across the Channel. He wanted all the Stukas and all the parachutists to be flung into an all-out attack on two southern airfields, Manston and Hawkinge. With those in German hands and operational for Me 109s and Stukas, German army units, ferried in Ju 52s, would take Dover and Folkestone.

These *points d'appui* would be supplied and supported from the air until a break-out could be made to take more of the English coastline. Milch's proposal was not accepted. Had it been, at that particular fulcrum in history . . . who knows?

5. It has been shown since the war that during the Battle of Britain whereas RAF claims of German aircraft shot down were twice too great, approximately, German claims were *five times* too great. These, on both sides were 'official' claims, and were regarded as propaganda. But whereas the RAF, fighting over its own land, was hardly influenced by the claims, the Luftwaffe was. An important difference between the two fine services was to be found in the quality of their operational intelligence. In this section the RAF was exceptionally strong, the Luftwaffe weak. British Intelligence officers were placed right through the service, down to squadrons, whereas their German equivalents were not seen in headquarters lowlier than those of Fliegerkorps until the end of the war was at hand. From 1940 on, RAF operational intelligence was remarkably well informed about Luftwaffe strengths, plans, aircraft, and equipment. As to claims of aircraft shot down, Air Intelligence during the Battle of Britain was unpopular with Fighter Command because it stated that the Command's claims were fifty per cent too high.

# 7

# Secret Roads in the Sky

*1940*

No battle had ever been so public. The bombs, the bullets, the spent cases, the burning aircraft, the pilots, German and British, living and dead, fell among the people. One doubts whether the Royal Air Force could have fought so well had it not had such constant and warm support from the whole population, from the other services, from the Government, from factory and inn, field, hospital, and shop. It was the country's battle for survival; it was the people's affair. How could the Germans ever have expected to win? (But then the Germans did not know about Watson-Watt and Tizard; neither did the people.)

Increasingly as the daylight battle continued, the enemy droned and rumbled over by night. It is hard to be courageous at night, and the wail of the sirens, the harsh noise of the guns, the whistle and crump of bombs made it still harder. The people fought to get used to these hateful and unnatural things. Only a handful of them in high places knew that Britain had entered the war quite unready for such night attacks. A secret battle was fought, and never made public, to protect the country from German night bombing of great potential menace.

* * *

Flight-Lieutenant H. E. Bufton was suddenly posted from his bomber squadron to the Experimental (Wireless and Electrical) Flight at Boscombe Down.[1] He was an experienced blind-approach instructor, using the Lorenz equipment which was common both to the RAF and to the Luftwaffe. He knew he had been called to Boscombe for something special, but he felt he had come down in the world when he was assigned an Anson training aircraft into which had been fitted an American short-wave receiver of the type used by the Chicago Police. Nineteen days before the air attack on England began Bufton was called to the telephone. Squadron-Leader Blucke, who had once been his blind-approach instructor, told him he had been brought there to hunt for a beam.

What sort of beam?

A beam sent over England from you-know-where.

No kidding!

He was to hunt for it on 30 and 31·5 megacycles, and geographically over the area between Huntingdon and Lincoln. Bufton mentioned the limited range of the Anson, and was told he might fly to Wyton, Huntingdonshire, refuel there, and take off in the late evening. If the beam existed it was perhaps switched off through the daylight hours.

That was a lovely summer night, the shortest night of a fateful year for England and for Germany. Bufton's Anson climbed slowly—it was never an exciting aircraft—to between four and five thousand feet.

There Bufton and his companion, Corporal Mackie, heard clear signals, dots on 31·5 mc/s.

He knew exactly what to look for. He brought the aircraft's funny little nose to the north. The night sky was empty. No glimmer from the factories below, working round the clock. . . . *Still dots*. . . . After a few minutes, flying they crossed what he was looking for—a beam, or continuous note. Then, as he again expected, they were flying through a zone of dashes. The old Anson was no longer dull. *Still dashes*, then no signal. Right! His job now was to plot the beam and if possible estimate its width. And up there he understood that he was charting what amounted to a secret road leading from Germany to the blacked-out multiple heart of industrial England. Beneath him were the machines and the skilled workers the Germans would wish to destroy.

He came down for petrol at Wyton, flew on immediately to Boscombe Down, and hurried to the telephone. He had orders to call Blucke, no matter what the hour. Blucke merely told him to get some sleep and to be at the Air Ministry in London early next morning. Then Squadron-Leader Blucke telephoned a young man who spent his days at the Air Ministry but who then slept in Richmond, a Dr R. V. Jones.

*   *   *

We last heard of R. V. Jones working successfully on infra-red methods for detecting aircraft, working in the Clarendon Laboratory at Oxford.

Sir Robert Watson-Watt, CB, LLD, FRS, the pioneer of Radar in Britain. (*Imperial War Museum*)

Professor Lindemann (left), Sir Winston Churchill's Scientific Adviser. To his left hand stand Air Chief Marshal Sir Charles Portal, Admiral Sir Dudley Pound, Sir Winston Churchill, and Major-General Loch. (*Imperial War Museum*)

*Left* A. P. Rowe, CBE, presiding genius of the Telecommunications Research Establishment. (Over his shoulder, his opponent, Field-Marshal Göring). (*A. P. Rowe*) *Right* General Kammhuber, creator of the Kammhuber Line which proved so costly to Allied bombers. (*Alfred Price*)

Professor R. V. Jones, CB, CBE, FRS, 'the tenth man' behind Britain's Radar war effort. (*R. V. Jones*)

Technically speaking, he had been attached to the Air Ministry Staff from 1936, and both Lindemann and Tizard had taken measure of his powers. He did a spell of duty in the Air Ministry from April to July, 1938, and during that short period A. E. Woodward-Nutt (at that time Secretary of the vital Tizard Committee) noticed that Jones 'continually got himself mixed up with Intelligence matters', and appeared to have a flair. Woodward-Nutt also had worked on research (at the Air Defence Experimental Department, Farnborough) with Jones, and knew him well. It was at his suggestion, strongly supported by Tizard, that Jones was brought back to the Air Ministry, and allocated a small office, but no secretary, as 'a scientist with a special interest in German weapons'. His beginnings afford a contrast with the more comfortable and apparently more assured ones of Major Schmid in Berlin. Those who have studied the air war in any detail know that Jones's appointment was a crucial one. At first it did not seem so to any but Jones himself. The terms of his employment were scarcely defined, and nine men out of ten would have failed to bring the new job to life.

Jones was the tenth man. He had been much influenced by his parents, who were deeply suspicious of German intentions after the 1914–18 war. 'Also relevant,' he says 'were the views of my headmaster at Alleyn's on the theory of forgiveness. He held that for a sinner to be forgiven, it was necessary that he should first repent. Since the Germans had never expressed repentance for the First World War, but only sorrow and dismay *that they had lost it*, they could not be forgiven for it. "And mark my words," he would say, "as soon as they are ready they'll be at it again." So I was well alerted, even before going to Oxford.'[2]

To all his searches and his advice—'the test of a good Intelligence service in war is not merely that you are right, but that you persuade your operational or research staff to take the right countermeasures'—he brought the logic of the scientist and, even more important, his honesty. Frequently Jones was to lay before his Ministry, the War Cabinet, the defence scientists, disagreeable evidence that they at first refused to credit. 'The path of truthful duty is not easy; there were several attempts to get me removed from my post because of my insistence on unpalatable facts being faced. I survived—but I might not have done so had the situation not been so serious.'[3] And as his deductions and discoveries proved to be accurate, his power grew.

'My first Intelligence task in 1939', Jones says,[4] 'was to report on the "secret weapon" which Hitler was alleged to have vaunted in an early war speech. After assessing the evidence and examining what he had actually said (rather than what he had been reported to have said), I concluded that he was not referring to a special weapon. I had just written my report (which was due to go to the Prime Minister, Mr Chamberlain) but had not circulated it, when the news of the magnetic mine broke. A report from Naval Intelligence came in that a German naval officer had said this was indeed the "secret weapon". It was tempting to alter my report. But I decided that this one remark ought not to be equated in weight to all my previous analysis, and so I rejected it—as it turned out, correctly. . . . I have seen committees, even up to cabinet level, almost stampeded by isolated pieces of "stop press" information, and there is no doubt that timing is an important device in the art of the advocate. But it should be eschewed in analysis proper.'

Jones's next problem was the Oslo Report, as it has come to be called, whose provenance is still a well-kept secret.

'At times of alarm, such as followed the outbreak of war and Hitler's speech, informers crop up in large numbers,' Jones told the Royal United Services Institution on February 19, 1947. 'Much of this information is useless, but soon after Hitler's speech one casual source came up whose information was of remarkable interest. It happened this way. Our naval attaché in Oslo received an anonymous letter telling him that if we would like a report on German technical developments all we need do was to alter the preamble on our German news broadcast on a certain evening, so as to say, "Hier ist London," instead of whatever we usually said. We duly altered the preamble.'

On November 4, 1939, the naval attaché found on his desk a packet of documents in German purporting to describe the new weapons at the disposal of the Reich. The sender described himself as 'a friendly German scientist'.

'He told us', Jones said 'that the Germans had two kinds of Radar equipment, that large rockets were being developed, that there was an important experimental station at Peenemunde, and that rocket-propelled gliding bombs were being tried there. There was also other information—so much of it in fact that many of our people argued it must be a plant, because no one man could possibly have known of all the developments that the report described. But as the war progressed, and one development after another made its appearance, it was obvious that the report was largely correct; and in the few dull moments of the war I used to look up the Oslo Report to see what was coming along next.'

In December, 1939, Jones drew up a report of his own. It suggested a drastic change in the structure of Intelligence, particularly scientific Intelligence. He wanted a single body, directed by a single man, and advising all three services. 'But the importance of scientific Intelligence was not yet appreciated. I failed to get an inter-service organisation. I also failed to get any help at all, even a secretary.' His effort to get a more logical framework having been rejected, he decided that, 'I would go on alone, to see whether I could prove my beliefs by practical demonstration. . . . The demonstration came even sooner than I had expected.'

* * *

As 1940 began, and what was popularly called the phoney war continued, Jones came to believe that the Germans had a system using radio beams with which they hoped to bomb accurately at night. In his searches of German material he had come across the word *Knickebein* (meaning a crooked leg). The Germans were ridiculously informative with their code names. *Knickebein* did sound like a description of a beam, or its emitter. Making his interest known, through the Directorate of Intelligence, he waited and watched until March, when a German reconnaissance aircraft, a Heinkel 111 of KG 26, was shot down near Scapa Flow. The navigator's notes were soon with Jones. He read:

*NAVIGATION*
*Radio beacons working on Beacon Plan A.*
*Additionally from 0600 Beacon Dunhen.*
*Light beacon after dark.*
*Radio beacon Knickebein from 0600 on 315.*

RAF interrogation of the comparatively few German airmen captured in those early days over British territory was both skilful and sensitive. Soon after Jones read the scribbles translated above a German prisoner said that the beam sent out by a *Knickebein* was so exact that two of them could pinpoint a London target with an accuracy of less than a thousand metres. He said, further, that *Knickebein* in some ways resembled *X-Gerät* (obviously taking it for granted that the British were familiar with both).

When another He 111 of KG 26 was brought down a diary was taken from the wreckage and flown to London. Jones read:

> *March 5: Two-thirds of the Flight on leave. Afternoon training on Knickebein, collapsible boats, etc.*

Going back through the files, he found that a He 111 had force-landed near Edinburgh soon after the beginning of the war, in October, 1939. He obtained the complete report on the examination of the Heinkel by Farnborough experts. They had put it on record that the Lorenz blind-landing set was many times more sensitive than its RAF equivalent. If German aircraft picked up *Knickebein* beams sent from Germany on their extra-sensitive Lorenz receivers, that would bear out Woodward-Nutt's theories at Farnborough. He had maintained that a radio transmission could be condensed into an exceptionally narrow, long-reaching beam.

Young Dr Jones understood his own position. Had he had the power he would have ordered up aircraft with listening sets to hunt for *Knickebein* beams. But he was new and junior. He must, as he said later, not bark until he knew his bark would be accepted as a danger signal. And he held two keys to the power chamber, since both Tizard and Lindemann knew him and, he believed, liked him.

Lindemann sent for him on June 12. The Prof wanted to know if Jones believed that the Germans had Radar. Jones switched to the possibility of the Germans having radio beam-aids for their night bombers. Lindemann argued against it. He said that all known evidence showed that radio waves on 30 mc/s frequency or thereabouts (Jones had taken the frequency from the captured notes and the Edinburgh Heinkel's Lorenz receiver) travelled in straight lines through space rather than curving with the earth's surface. It would be impossible therefore, for *Knickebein* beams to reach the Midlands (even to aircraft flying at twenty thousand feet) let alone to reach Scapa Flow . . . . But next day Jones called again on Lindemann. He had found an unpublished paper by Thomas Eckersley, a scientific adviser to the Marconi Company, and a respected expert at the Air Ministry. From Eckersley's graphs it seemed that radio beams from Germany could be received by aircraft over much of Britain. The Prof had had Jones's former evidence in his head all night. He now sat down and wrote an URGENT memo to Churchill:

> There seems to be reason to suppose that the Germans have some type of radio device with which they hope to find their targets. . . . It is vital to investigate and to discover what the wavelength is. If we knew this we could devise means to mislead them. . . . If they use a sharp beam this could be made ineffective. . . .

Churchill wrote across the bottom of the Prof's memo, 'This seems to be intriguing, and I hope you will have it thoroughly examined.' He had it taken to the Secretary of State for Air, Sir Archibald Sinclair, who next day, Friday, June 14, asked Air Marshal Joubert to form a committee of enquiry. The following day the committee met and did exactly what Jones would have done had he (then) had the authority. They had put Squadron-Leader R. S. Blucke (who had done the first Radar-proving flight in the Heyford) in charge of the flying side of the investigation. Three Ansons were to be fitted with suitable receivers and experienced Lorenz-trained pilots were to be chosen by Blucke. . . . *As the committee sat, the German army was marching into Paris.*

Another German airman had now spoken of *Knickebein* under interrogation, and had confirmed that *Knickebein* beams were picked up on the adapted Lorenz receivers in German aircraft. And on Tuesday, June 18, papers from a German aircraft shot down weeks earlier in France were on Jones's desk. In them he found:

| *Long-range Radio Beacon* = | *VHF* |
|---|---|
| *1st Knickebein* | *54 39* |
| | *8 57* |
| *2nd Knickebein* | *51 47 5* |
| | *6 6* |

Transferring the *Knickebein* positions to the map of Germany, he found the first was at Bredstedt and the second near Kleve. Scapa Flow's bearing from the Bredstedt position was 315°, so that tallied. The two positions were sufficiently far apart to give reasonable cross-bearings or 'fixes' on most important targets in Britain.

Next Jones got the notes of a dead wireless operator. A minelaying

Heinkel of KG 4 had been shot down. At the head of his complete (roneo) list of German radio beacons, with their frequencies, the dead man had written '*Knickebein*, Kleve, 31·5'. The RAF monitoring service confirmed that all the frequencies for the beacons were correct for the night concerned.

And finally, on the morning of June 20, a He 111 with twin Jumo 211 motors, a new type, was winged by a fighter over South-Eastern England. The wireless operator landed by parachute. Before he was found he tore his working notes into very small pieces and these, stout fellow, he was burying when they came upon him. An astute RAF Intelligence NCO saved every scrap of paper and by three the next morning these had been gummed together and were on their way to London. Jones read:

| | | | | | |
|---|---|---|---|---|---|
| *VHF* | *54* | *38* | *7"* | *North* | *Stollberg* |
| *Knicke* | *8* | *56* | *8"* | *East* | |
| | *51* | | | *N* | *(30 mc/s)* |
| | *1* | *30'* | | *Eqms* | |
| *Cleve* | *51* | *47'* | *4"* | *N* | |
| | *6* | *2'* | | *E* | |
| | *55* | | | *N* | *(31·5 mc/s)* |
| | *2* | | | *Eqms* | |

This timely jigsaw confirmed for Jones the positions of two *Knickebein* stations. That the last positions were given more accurately than the earlier ones and that Kleve was spelled in them with the old-fashioned C was a better Intelligence confirmation than conformity. The two other positions were out in the North Sea, possibly turning points.

Later *that same morning*, June 21, Jones returned to his office in the Air Ministry building and found a note asking him to report immediately at the Cabinet Room, 10 Downing Street. He got through to the Ministry's executive department. It was genuine! They had been hunting for him everywhere. Grabbing his papers, he tore round to Downing Street. He was appallingly late. The special meeting called by Mr Churchill had been in progress for half an hour. The door was opened for him and he hurried in, making his apologies to the Prime Minister.

Churchill sat at one side of the table, with Professor Lindemann on his left hand and Lord Beaverbrook, the owner of Express Newspapers and the new Minister of Aircraft Production, on his right. Facing the redoubtable trio were Sir Archibald Sinclair (the extraordinarily handsome Secretary of State for Air), Sir Cyril Newall (Chief of Air Staff), Tizard (Newall's scientific adviser), Watson-Watt (now representing Communications at the Air Ministry), and Portal and Dowding (Commanders in Chief Bomber and Fighter Commands). The meeting was obviously urgent and tense, and seemed the more so because the proceedings were secret. No secretary was present and no minutes were taken.

As Jones entered the handsome room Lindemann beckoned to him to sit among the gods at the emptier side of the table. But the young man hesitated. He would not take sides. Tizard was as much a friend as was Lindemann, and there was hostility in the air. He sat down alone, at the end of the table nearest the door.

An argument was continuing. . . . Did the beams exist or didn't they? Tizard was sceptical. The Prime Minister turned to look down the table. He asked Jones a technical question and Jones, feeling he could not sensibly answer it out of context said, 'Hadn't I better tell you the story from the beginning, sir?'

He laid before them the evidence from German aircraft and air crew given above, and mentioned, as he had previously done to Lindemann alone, the deductions of Eckersley, the Marconi scientist. He convinced the meeting. And Churchill, marvelling at the size and vitality of the speaker, as well as at his fresh face and extreme youth, found a jingle from *The Ingoldsby Legends* running through his head. . . .

*But now one Mr Jones*
*Comes forth and depones*
*That, fifteen years since, he had heard certain groans*
*On his way to Stonehenge (to examine the stones*
*Described in a work of the late Sir John Soane's),*
*That he'd followed the moans,*
*And led by their tones,*
*Found a raven a-picking a drummer boy's bones.*[5]

The meeting at Number 10 was both good and bad for the country. In one way it was providential, since it ensured maximum effort against the German devices. In another way it was harmful. For, certainly without intending anything of the sort, Jones by his evidence had wounded, at any rate in Churchill's view, the prestige of the country's finest defence scientist. Tizard had been dubious in his assessment of the possibility of such beams. (So, after all, had Lindemann until Jones showed him Eckersley's graphs.) We have seen how vital Tizard was in the development of Radar, and at this moment in 1940 his potential as a war winner was still boundless. At the Prime Minister's shoulder sat the subtle and implacable Lindemann. And whatever exactly transpired in the unrecorded meeting, Tizard decided that after it his position as Scientific Adviser to the Chief of Air Staff was impossible. He went to his club, the Athenaeum, and in that high and formal edifice, austere and rather dusty with the wartime shortages of staff, he wrote a letter of resignation. He carried the letter to his chief, Sir Cyril Newall, that same evening, and Newall, appearing to agree with him, accepted it. . . . In less than a month Sir Henry Tizard was asked to lead the vastly important scientific mission to the United States that he had been advocating for nine months. He carried with him great gifts in the form of British secrets, including that of the war winner, the cavity magnetron valve.[6] It was an honourable task, and nobody could have been more worthy of it than he; yet most British scientists, had they known him to be out of the country, would have felt the weaker for his absence. And when he returned his services were parsimoniously used.

On leaving the meeting at Number 10, Jones returned to the Air Ministry where, in the office of the Director of Signals, Air-Commodore Nutting, he happened to meet the Marconi expert, Eckersley. That morning, basing his evidence partly on Eckersley's paper, he had convinced the country's leaders that the *Knickebein* danger existed. *But Eckersley himself now doubted.* Then, what about that series of graphs? Jones asked. Oh, those graphs! Eckersley renounced them. He now felt that he had been stretching theory. He now honestly doubted if radio signals on the 30 megacycles band would curve round the earth.

While Jones was ruefully considering this reversal, and balancing it against his other evidence, the significance of Eckersley's denial had not been lost on the others present. Indeed the Deputy Director of Signals, O. G. W. Lywood, had his hand on the telephone. He thought that Blucke's third flight in search of the German beam could be cancelled.

'Although I was shaken by Eckersley's statement,' Jones says, 'I staked everything on the flight. I told Lywood that the Prime Minister had given orders for the flight, and that if it were cancelled I would see that he knew who had stopped it. I then went home and spent one of the most miserable nights of my life. . . .'[7]

As he left the Air Ministry to catch a train to Richmond, Bufton and Mackie were flying from Boscombe Down to Wyton, prior to the start of their night's search.

## NOTES TO CHAPTER 7

1. Clark, *The Rise of the Boffins*, pp. 110–12.
2. Jones, letter to the author.
3. Jones, *Minerva*, Vol. X, No. 3, p. 446.
4. Jones, *Minerva*, same No., p. 449.
5. Churchill, *The Second World War*, Vol. 2, p. 339.
6. The valve that defeated the U-boats, see Chapter 11.
7. Jones, letter to the author.

# 8

# Rémy and Pol

*January, 1942*

During his listening watch in Paris on January 24, 1942, Bob (Robert Delattre) took down two messages from headquarters in London. That evening in a rented flat in the Avenue de La Motte-Picquet, Bob's chief, the legendary Rémy, decoded the messages with the help of his wife Edith. Open on the table before them lay Michelin Map No. 52, which placed in its long rectangle the towns of Le Havre, Rouen, Beauvais, Abbeville, and Amiens. It was rare for an important agent such as Rémy to decode messages; but Rémy was an unusual man, a man with a lot of the boy in him, and furthermore he, like his wife, adored codes. At last they read[1]:

> 24.1.42 TO RAYMOND CODE A NO 49
> need information indicated questionnaire message that follows stop inform us within forty-eight hours delay necessary obtain this information observing following conditions firstly do not act yourself nor gravely risk members your organisation secondly do not compromise success operation julie stop to deceive boches in event your agent taken he be ready to reply same question not only for place chosen but for three or four other similar places on coast stop to follow. . . .

Rémy was addressed as Raymond in W/T messages. Operation Julie was the code name for the Lysander pick-up of Rémy himself in the next full-moon period. He was wanted in London for consultation with 'Passy' (Major André Dewawrin) who commanded the BCRA (Bureau de Contre-Espionage, de Renseignement, et d'Action) of General de Gaulle. The continuation message was:

24.1.42 TO RAYMOND CODE A NO 50
questionnaire firstly position and number machine-guns defending cliff road at theuville repeat theuville on coast between cap antifer and saint jouin latter being seventeen kilometres north le havre secondly what other defences thirdly number and state preparedness defenders stop are they on qui vive stop firstclass troops or old men stop fourthly where are they quartered fifthly existence and positions barbed wire ends

* * *

Rémy (Gilbert Renault) was one of the most remarkable of the Frenchmen who joined de Gaulle when France fell to the Germans. The odd thing was that at the outset of war the French Army, to Rémy's fury, had rejected his services as a combatant because of his large family and the wartime importance of his civilian work.

When Paris fell he and his family were staying with his mother at Vannes. A fortnight earlier Edith had told him that she was carrying his fifth child.

'Better die than live as a lackey of the Germans,' Rémy said. 'Don't you agree? We are most fortunately placed here—near the sea. It need only be a question of going to Lorient, where I can certainly get a boat either to England or to North Africa. What do you say, my love?'

'Go.'

'If you say, stay, dearest one, I won't go.'

'No, go. You must do your duty as you see it.'

'It would be ignoble not to join de Gaulle at this juncture.'

'My loved one! I agree.'

He took with him his youngest brother, Claude. Five days after leaving Vannes, the pair landed from a fishing boat at Falmouth and at once made for London, and General de Gaulle's headquarters in St Stephen's House.

*Now* Rémy intended to get his own back for those dreadful months at the beginning of the war when they had refused to let him *do* anything. He at once volunteered for a Secret Mission, and he says that he was accepted because his passport was covered with Spanish visas dating from a period of his life when he had been preparing to make a film about Christopher Columbus. Meanwhile he had the happiness of seeing Claude commissioned in the Gaullist forces. Rémy quickly went back to France through Portugal and Spain. His selection was a brilliant one, even by Passy's standards, because he was a genuine and talented person, and far better, he had the luck of the devil.

A modest man, supremely friendly, whose religion was the mainspring for his every calculation, decision, act, Rémy thus describes his beginning as an agent: 'I would never have been able to carry out this assignment in a foreign country or for a less righteous cause. But I was on my own soil, among Frenchmen with whom the enemy could not ingratiate himself, and whom he could not intimidate. . . . From the day when Jean Fleuret, former union leader of the Pilots of the Port of Bordeaux had agreed to work with me' (note that Rémy says with rather than for) 'I realised that my task would be easy. I only had to find, in the ports, the railways, the factories, the administrations, men and women of goodwill—and I would receive a mass of valuable information which would surpass all hopes. So, mesh by mesh, was woven my network. My part consisted simply in convincing people who were eager to be convinced for their country's sake, then in holding all together. *Naturally, my family were among the first to join our ranks. . . .*' Rémy, indeed, seemed able to break almost all the rules of espionage. And how his brave family suffered! After a long imprisonment in Fresnes, Romainville, and Compiègne, his mother and his sisters Hélène, Jacqueline, and Madeleine were freed, but his sisters May and Isabelle were deported to the female concentration camp of Ravensbrück. His brother Philippe, also deported by the Germans, was killed in Lübeck harbour a few hours before the British Army arrived there.

Convinced that he and his followers were a brotherhood of man favoured by God, Rémy called his group the CND (Confrérie Notre-Dame). When he had left London General de Gaulle had taken him by the hand and had said firmly, 'Au revoir, Raymond, je compte sur vous.' It was a tall order, and the tall general must have known it, since Rémy had been given the whole Atlantic side of France to cover, from Hendaye up to Brest. He had accomplished wonders in that enormous area for a twelve-month when the equally important Réseau Saint-Jacques, run by his friend Maurice Duclos, and covering the North Coast of France, from Brest to Dunkirk, was infiltrated and totally destroyed. From London Passy asked CND to take over the information services made void. He began to build again, simplifying all moves and problems as only clever people can. He describes his task at that time of danger as 'putting living tile upon living tile'. Some of his 'tiles' were sent from London (such a one was Bob, a Frenchman who had been trained in the schools of SOE[2]). Most were recruited in their home areas from the professional classes (many were architects), from the officer class, from the aristocracy and the bourgeoisie, from the peasants, and from those in 'commerce'.

One of his recruits in the northern area taken over from Duclos was an officer of the Armée de l'Air. His name was Roger Dumont and (thinking of Pol Roger champagne) 'I gave him the code-name Pol and told him he would lead our Luftwaffe section.' Shortly after taking over this work, Pol had laid before Rémy a detailed report from his friend Roger Hérissé which described closely guarded German radio installations just north of Bruneval.

Pol was in Paris on January 24. Rémy soon got hold of him, and when they were alone watched his dark head bent over the two 'Theuville' messages and over the map. The text of the second message knocked out the theory of air attack. No, 'they' were thinking in terms of a raid either with commandos landed from the sea or with parachutists. What was there at Bruneval or Theuville that was so important to 'them' in England? Thinking this over, Rémy was of the opinion that nobody should attempt to get near the actual German installations.

'The best way to be discreet,' he said to Pol, 'is not to begin to try to understand these two messages. You agree with me?'

'Certainly.'

'How long will it take you?'

'Ask them to give me a fortnight, to get everything properly buttoned up.'

'Watch yourself now, my dear fellow. This is an operation that must not be prejudiced by rashness. I know I can trust your discretion. We'll get Charlemagne to give you a hand. That's right in his back yard, Bruneval, and he's very very bright.'

### NOTES TO CHAPTER 8

1. Colonel Gilbert Renault (Rémy) is the authority for most of this and subsequent chapters concerning his end of the business. Much of the material here used can be found in his *Bruneval: operation coup de croc.* That book and

his others on the Resistance as he knew it (see Bibliography) give a direct and unforgettable picture of the Resistance with its agonies and its flashes of pleasure and wonder. Here the author feels himself at home. It is to people like those who worked with Rémy that he owes his life.

2. SOE, Special Operations Executive, was divided into two sections so far as operations in France were concerned. This writer served briefly in the British section, Rémy saw long service in the Gaullist one. See Foot, *SOE in France*.

# 9

# Doctor Plendl's Devices

*1940/41*

If it went to work at night Bomber Command had been expected to find its targets by dead reckoning and astral navigation, an uncertain procedure. Flight-Lieutenant Bufton therefore administered a profound shock when, on the morning of June 22, 1940, at the Air Ministry he described to very senior officers and defence scientists the clarity, power, and narrowness of the German beam (see p. 64).

Churchill refused to be dismayed, and he had at his elbow the scientist-confidant who assured him that any radio beam sent from two hundred and sixty miles away was itself vulnerable. The first thing to do was to create a defence unit with complete civil and military authority, facilities, aircraft, and all the scientific help it could use. The unit, 80 Wing, established itself at Garston Aerodrome, and under Wing-Commander E. B. Addison set about a defence task that, with some frankness, was code-named HEADACHE.

Short-wave sets established on top of the Chain Home towers could pick up the German beams. The wretched occupants of those alarming crows' nests were connected by telephone with our friend R. S. Blucke, now climbing in rank as his duties multiplied. He sat in the centre of the web, Fighter Command Headquarters, Bentley Priory. 'When about dusk the German beams were switched on, the men on the towers would be able to pick them up and let us know, for instance, if a beam was going between Tower A and Tower B,' Blucke has explained.[1] 'That would give us a clue to the beam's position, and one of our chaps would go up in an Anson and fly back and forth until he picked up the beam, which could then be plotted.'

After the fall of France British Intelligence had been caught in a vacuum from which it instantly began to emerge when it became clear to Europe that the war was not yet lost. Within three weeks of Bufton's discovery-flight Dr Jones knew about *Knickebein* stations near

Cherbourg and Calais; a third, near Dieppe, began testing on August 23, 1940.

A first solution might have been bombing; but although the aerials were massive, the transmitters were the smallest of targets, and bombing then, as now, was by its very nature imprecise. As to defence, electro-diathermy sets used in hospitals to cauterise wounds were requisitioned and altered to transmit on *Knickebein* frequencies. They were then installed in police stations where twenty-four hour watches were kept. The policemen only switched on when asked to do so by 80 Wing. Then Lorenz blind-approach transmitters were modified and strategically placed. They were thought to have a certain usefulness in distorting the beams over a short range. In addition 80 Wing set up spoofing beacons, or 'Meacons'. The Luftwaffe, in Germany, the Low Countries, France, and Norway, could navigate on more than eighty radio beacons. The Meacons, strategically sited in Britain, and set daily to German frequencies, were a complication for Luftwaffe navigators.

Something more positively effective, however, had to be designed and built in quantity. This problem was passed to the Research Establishment formerly at Bawdsey and now in Dorset. A team led by one of the Establishment's outstanding young scientists, Dr Robert Cockburn, worked on a *Knickebein* jammer called 'Aspirin'. Cockburn was destined to be an outstanding figure in the radio war now beginning. He had been a science master at West Ham Municipal College until, in 1937, he was persuaded to join the radio staff at Farnborough to help develop Fighter Command's VHF equipment.

Imagine a *Knickebein* as two transmitters side by side, the right hand one sending out a long beam of dots, the other a beam of dashes. In the middle, where they narrowly overlap, is the beam aimed at the target, and in this beam the dots exactly key in with the dashes to give a continuous note. (It helps to understand this if one thinks of square dots and rectangular dashes.) Aspirin could transmit on any *Knickebein* frequencies, but it churned out only dashes. They flooded the German beam, because Aspirin was powerful. When the German flew into his own dash zone he would veer to find the central beam, but in it he would still get (Cockburn's) dashes. So he would go on turning until, in the dot zone, he got a tangle of dots and (Cockburn's) dashes that occasionally synchronised into a false beam note. Cockburn had worked out a system for bending the *Knickebein* beams. But the Aspirins proved so successful that they were set up to guard most important targets.

*Left* Major John Frost, now Major-General J. D. Frost, CB, DSO, MC, who led the Bruneval Raid. (*Imperial War Museum*)
*Right* Lieutenant Dennis Vernon, the Sapper officer who led the dismantling party. (*Dennis Vernon*)

Flight-Sergeant Charles Cox (*right*), the Radar expert who stole the *Würzburg*; pictured at a Bruneval commemoration ceremony, flanked by a parachute sergeant. (*C. W. H. Cox*)

*Above* Britain's round-the-clock sentinels: a typical Chain Home Radar station. (*Imperial War Museum*)
*Below* A *Würzburg* Radar abandoned by the Germans; this was used to provide fire-control data for a flak battery, and displays a claim of three aircraft shot down. (*Official U.S. Air Force Photo; Alfred Price*)

Captured Luftwaffe officers and men began to complain of the 'unreliability' of the beams that had formerly kept them on target. Crews who relied on *Knickebein*, they said, would find themselves flying round in small circles.

By accident or by Aspirin a few bombs fell on London. In reply Churchill asked for a night raid on Berlin. And then Hitler and Göring could not be held. It was tempting for them to imagine that Britain might be defeated by a continuous and massive bombing of the capital. The decision to martyrise London by day and night bombing meant the end of their chance of winning the Battle of Britain because it relieved the pressure on Fighter Command, which daily grew stronger while the Luftwaffe got relatively weaker. The first daylight attacks cost the Germans so dear that the continuation had to be confined to the dark hours. Serious night bombing of London began on September 7, 1940, and until November 13 (with the exception of only one night of storm) an average of one hundred and sixty bombers attacked the capital in every twenty-four hours. London, near the coast, and with the silver snake of the Thames laid across its vitals, was the easiest of targets to hit in any weather. The town played its full part in the defeat of Germany not only because of its spirit but also because of its enormous spread. Had the *Knickebein* square system worked as it was meant to—or had it not been interfered with—each night that number of bombers should have put into a selected target area of the town *a bomb every seventeen yards*. That would have been lethal . . . It makes one wonder how many Londoners have heard of Robert Cockburn and of R. V. Jones.

Before the *Knickebein* threat had been settled Jones was pre-occupied with *X-Gerät*, the X-Apparatus, whose code name in the Air Ministry files was RUFFIAN. He had learned that *X-Gerät* receivers were fitted only in the Heinkel 111 bombers of a single, élite, independent squadron, Kampfgruppe 100. KGr 100 had been brought earlier than some of Göring's advisers thought wise into the Battle of Britain. On the night of Eagle Day half the squadron (twenty Heinkels) had performed a meticulous exercise against a Spitfire shadow factory on the outskirts of Birmingham. Eleven bombs middled on the factory, and the remainder were very close indeed. (Thanks to Jones's department RAF monitors were familiar with KGr 100's call sign—6N.)

Shortly afterwards a radio-search aircraft of 80 Wing piloted by none other than H. E. Bufton (odd how two practical flying men,

Bufton and Blucke, keep turning up in this story) was doing a routine flight off the Cherbourg Peninsula—*not* usually a comfortable locality. 'We had one of the first centimetric receivers, but we also had one of the old 100 mc/s sets,' Bufton said later.[2] 'On that set we found ourselves in a maze of signals on the 70 mc/s band, apparently radiating from several transmitters . . .' The pilot who had found the *Knickebein* beam had also found the *X-Gerät* multiple beam.

What was *X-Gerät*? As early as 1933 Dr Hans Plendl, a specialist in Herzian developments, began to experiment with an idea for a blind-bombing aid. Telefunken were working on the same problem, and they came up with *Knickebein*. Plendl saw that *Knickebein* would be interfered with by a skilful enemy. His own scheme was more complex, and it says much for German thoroughness that it was persevered with at a time when German air superiority was so marked. *X-Gerät* consisted of a parent beam that could be aimed at any chosen target, and three cross beams. The parent beam *Weser* was, as Bufton had discovered, transmitted from Cherbourg, while the three cross beams, *Rhein*, *Oder*, and *Elbe*, came from Calais. In order to achieve an exceptionally accurate target beam, Dr Plendl had surrounded it with a sheath of parallel beams. The aggregate formed a wide highway, which KGr 100 alone could follow.

In mid-September Lindemann minuted Churchill, described the activities of KGr 100 'stationed at Vannes; home station Lüneburg, and reserve station Köthen . . . Bombing accuracies of the order of twenty yards are expected. With the technique they are developing this does not seem impossible.' Lines of action, he suggested, might be to wipe out the squadron by a concerted attack; to bomb the beam stations, which would be all but impossible because they were almost invisible targets; to destroy the stations by special commando operations; or finally, to devise radio counter-measures.

Work was soon progressing day and night at the Research Establishment. The jammer devised for *X-Gerät* was code-named 'Bromide'.

Throughout September the X-beams were often warmed up during the day, and 80 Wing sometimes knew what orders the bombers at Vannes were getting for that night. Then, at the beginning of October, there was an alarming change in KGr 100's tactics. Instead of dropping their usual 250-kilos HE bombs, its Heinkels dropped incendiaries. The implication was that KGr 100 was going to lead the ordinary bomber squadrons, and mark out the target for them by lighting fires in its centre. This was the beginning of the Pathfinder system of bombing which the British were later to develop.

KGr 100 took off from Vannes on November 5 to bomb Birmingham. Early the following morning one of its Heinkels was hopelessly lost over Southern England. Beguiled by a Meacon operating near Bridport in Dorset, the pilot thought he saw the French coast ahead, and as he was all but out of fuel, attempted to put the aircraft down on the beach, which was the enormously long and spectacularly beautiful Chesil Beach, a steep bank of round pebbles. As the Heinkel came down at the edge of the surf and undertow one German was killed and two were injured. The wrecked machine was surrounded by armed British soldiers. They secured a rope round the fuselage, which displayed the letters 6N + BH and the Squadron's device, a Viking ship.

While the troops set about salvage, a Royal Navy inshore patrol vessel came on the scene and asked the Army what it thought it was doing. As the aircraft was in the water, salvage was a Navy matter. The Army was displeased and disappointed, but the Navy took the rope aboard and dragged the aircraft into deeper water prior to securing it to a derrick and hoisting away. The rope broke and the Heinkel vanished into the sea.

'It is a very great pity,' the Prof wrote to Churchill that same morning, 'that inter-service squabbles resulted in the loss of this machine, which is the first of its kind to come within our grasp.'

However, the *X-Gerät* receivers were recovered and were rushed to Farnborough where Jones and Cockburn were among those who examined them. As the examination proceeded it was realised that *X-Gerät* was very difficult to beat. It worked on five frequencies, and had three further stand-by frequencies available to evade jamming. And the eight frequencies were chosen each night from twenty that the system could use. In addition there were both radio- and audio-frequencies. Jones was always interested in the dated inspection stamps that were a feature of German military equipment with its excellent routine servicing. The earliest date he found on the receivers' stamps was pre-war, 1938. It was frightening to think that the Germans had possessed so sophisticated a device before the war. And extreme interest was aroused by the number of jamming safeguards built into the sets.

Jones had got wind of a German plan, code-named 'Moonlight

Sonata', to wipe out three English industrial cities beginning, weather permitting, on November 14/15. He thought that the targets would be Wolverhampton, Birmingham, and Coventry; but he did not know in what order the attacks would be made. The defences were short of time and jammers, and there was a hectic rush.

'We did not know on the morning of November 14 that the target was Coventry,' Jones says.[3] 'Addison telephoned me at six that evening, and neither of us, nor anyone else in England, knew what the target was to be. We *all* knew that something big was on, and there had been some wild guesses by members of the Air Staff and others which further muddled the issue.

'Addison appealed to me to interpret the results of our listening flight, so as to tell him which frequencies to set the jammers on. It was a nasty problem, with five beams to jam and only three jammers available. It could just be done, if we picked the three most vital beams and got their frequencies right. It happened that I had recently broken the coding system for the German frequencies, and knew that, with one exception, they must all be either whole numbers of megacycles or whole numbers plus a half between 66·5 and 75·0. None of the measurements Addision telephoned to me squared up with this, and so I had to guess. I gave him my guesses, and then went home.'

* * *

Taking off from Vannes after nightfall, the Heinkels of KGr 100 kept well on the outside edge of the wide *Weser* beam, to avoid British night fighters, which were beginning to be dangerous, especially in moon periods. They crossed the British coast near blacked-out Christchurch, the Priory showing clearly in the moonlight. Their next landmark was the soaring and so-English spire of Salisbury Cathedral. At Swindon, where the rails glinted dull silver in the marshalling yards, the Heinkels began to edge in for the bombing beam. They crossed the coarse *Rhein* beam at 1906 hrs. That was the signal for the pilots to close on the narrow *Weser* bombing beam, and hold to it, flying straight and level for twelve miles. The German aircrews heard strong interference on that night's frequencies, but they could just pick out their own signals. Each observer started his bombing clock as they swept through *Oder* cross-beam while the pilot concentrated on flying within the bombing beam. Three hundred and thirty

X-GERÄT TARGET COVENTRY
ON THE NIGHT OF NOVEMBER 14, 1940

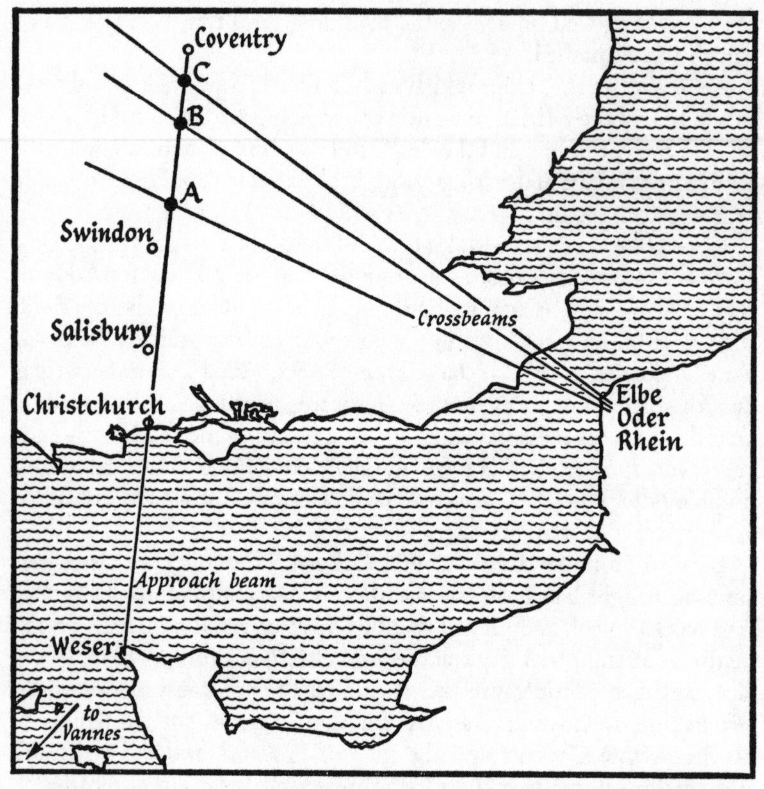


*A*—1906 hours. German pilots cross Rhein coarse beam and close on to narrow bombing beam (Weser).
*B*—1917 hours. Oder signal. Observers push first buttons to start bombing clocks.
*C*—1922.30 hours. Elbe signal. Observers push second buttons. Clocks' first hands stop; second hands start moving.
1924.20 hours. COVENTRY. Hands on bombing clocks overlap and firebombs automatically released.

seconds after *Oder* they flew into *Elbe* and each observer pressed the second button. The pilots continued straight and level while the two hands of the bombing clock in each Heinkel began to close. Not two minutes later the German incendiaries, directed from far-away Calais, fell in the centre of Coventry. Already the bomber squadrons were winging in for the kill. They came over Sussex and Hampshire and over the Wash. Bombing conditions were perfect and they had been well briefed on the main industrial targets, nearly all of which were smashed.

At dawn the sky, from which such horror had poured, was serene, while below the citizens and their ready helpers from outside strove to save the entombed and the desperately wounded, and to cope with threats of drainage in drinking water, broken gas mains, fires. Coventry was not dead—quite.

Jones's first thought on waking was of the night's events.[4] He quickly learned of Coventry's ordeal, and thought that his 'guesses' for Addison had been wrong. 'Actually,' he thought, 'it would have been a fluke if they had been right.' But in the post mortem that quickly followed Jones's 'guesses' proved to have been correct. What had gone wrong was that one man in the counter-measures organisation had made a mistake with his measurements, and as a result, while the British jammers were spot on in other respects, they were set to produce the wrong audible note which was lower than the German one—1500 instead of 2000 cycles. A costly mistake.

Bomber Command and the War Cabinet, examining the chaos that one short night had brought, saw that the German Pathfinder system had worked well, and judged that this form of aerial holocaust on the centre of an industrial city was the most effective form. What they possibly failed to consider was that soon after the holocaust the industrial production of Coventry *rose.* As for the Germans, their Intelligence services wrote Coventry off the map of England, and the Luftwaffe did not ram home near-destruction by a timed succession of attacks. *That* omission was noted by those in England who thought the war could be won by aerial bombing of Germany.

Next on the 'Moonlight Sonata' score was Birmingham. But now the audio-frequencies error in the jammers had been rectified, and the incendiaries and the bombs fell south of the target centre and in many cases outside the city boundaries. Jones had made great efforts with Anti-Aircraft Command to get major reinforcements of guns and searchlights round Wolverhampton. The defences waited, the guns silent, and Jones, had he been less confident in his own findings, would have felt embarrassed. But within three or four weeks confirmation came through that OKL had cancelled the Wolverhampton raid because a German day reconnaissance flight which was to have preceded it had shown the increase in the defences.

By the end of 1940 Plendl's *X-Gerät* had been controlled, indeed mastered, by 80 Wing and the scientists. But Jones and his new assistant, F. C. Frank, a young physical chemist, were on the track of yet another German blind-bombing device. In June Jones had learned that new 'Wotan' transmitters were being set up near Cherbourg and Brest. He deduced from the name Wotan (those Germans, with their pregnant code-names!), signifying Odin the one-eyed, that the new transmitters were to operate a single-beam system. Such a system, oddly enough, was the only one mentioned in the Oslo Report, where it was described with fair accuracy. (One curious aspect of this is that Plendl told R. V. Jones after the war that he had not even thought of that particular system until 1940.) Jones's deductions were correct, even if the 'one-eyed' lead was not accurate, for this was indeed Dr Plendl's second system, *Y-Gerät*. The German code-name for *X-Gerät* was Wotan 1, and that for *Y-Gerät* was Wotan 2. The British had two names for the latest menance, Wye and Benito.

At first they failed to analyse the new beam, and small wonder. General Martini had taken two whole hours to explain Benito to Göring, who at the end of it announced that he was completely flummoxed. He said it sounded too complicated, and in a way he was right.

When Cockburn put the beam on a cathode-ray tube he and his friends grasped its nature. The beam radiated three directional signals every second. The German bomber carried an electronic analyser by means of which certain emissions from the ground station were returned to it from the aircraft. The ground station could thus know the aircraft's exact distance away, along the beam, and when the aircraft was over the target the ground station told the bomb 'aimer' to release his bombs. It was a fantastic invention. It was possibly even more accurate than *X-Gerät*, and its operation needed only a single ground station; these could have proliferated along the French coast. But thanks to the Oslo Report and the strength of the scientific opposition, *Y-Gerät*'s usefulness was ending before it had been used—while it was still being worked up by the Heinkel 111s of KG 26 flying out of Poix, and using Wotan 2s at Poix and Cherbourg.

Cockburn named his *Y-Gerät* spoofer 'Domino'. His first Domino was operating at Highgate in North London in February, 1941, and very shortly the second was ready at Beacon Hill on Salisbury Plain. When the second one got going the Germans slashed in with a precision air attack on Beacon Hill that was nearly successful.

Dominos picked up the *Y-Gerät* ranging signal from the German bomber, and the BBC television transmitter at Alexandra Palace re-radiated the signal on the Wotan 2's frequency. This cancelled out the German system. There were feints and evasions. The Germans set up new ground stations and altered and re-altered frequencies. But only eighteen times were bombs dropped out of eighty-nine *Y-Gerät* bombing sorties.

Then, on the night of May 3, KG 26 lost three *Y-Gerät* Heinkels over England. When he analysed and tested the resulting mass of captured equipment, Dr Cockburn saw that, 'Unlocking the Wye-beam was a bit of cake. The Germans had fallen into the trap of making things automatic. All one had to do was to radiate a continuous note on the beam's frequency. This filled in the gaps between the signals, unlocked the beam analyser, and sent the whole thing haywire . . .'[5]

Cockburn had a new jammer, 'Benjamin', in operation on May 27. He knew that a single additional circuit in *Y-Gerät* would have protected it from the attentions of Benjamin. But the circuit was not fitted. And the Luftwaffe was being called away unit by unit, with increasing momentum—to Russia.

### NOTES TO CHAPTER 9

1. Clark, *The Rise of the Boffins*, p. 114.
2. Clark, *op. cit.*, p. 117.
3. Jones, letter to the author.
4. Jones, letter to the author.
5. Price, *Instruments of Darkness*, p. 49.

# 10
# A Black Market Luncheon

*January, 1942*

'Charlemagne', a CND agent in Le Havre, was a garage proprietor, Charles Chauveau by name. He came to Paris in a Simca 5 to pick up Pol at Rémy's request. As soon as the pair of them reached the border of Seine-et-Oise, Charlemagne drove into a wood and changed his false number plates for authentic ones corresponding with his permit to drive in Seine-Inférieure. The risks of such a journey at that time were great.

'Where are we making for?' Charlemagne asked.

'I have to study the coast and its approach roads in the neighbourhood of Saint Jouin and of Bruneval. Can you help there?'

'Of course, my friend. I know those corners like my own pocket.'

Charlemagne took rooms in a Le Havre hotel where no questions would be asked, and no identity cards would be needed. The sheets were damp, and it was so cold in the unheated tenement that Pol could not remain in bed. He spent the night shivering on a hard chair, fully dressed. In the morning Charlemagne borrowed two wheels with chains for the Simca. He had been warned that the steep roads round Bruneval had thirty centimetres of snow on them—over ice.

It was wiser for Frenchmen in those days to keep off their own main roads, particularly in that heavily garrisoned Channel area. As soon as he could, Charlemagne left the Le Havre–Etretat road, the N 40, and drove through Heuqueville and Saint Jouin. Another two and a half kilometres and they were at the Calvary road junction at the eastern end of Bruneval hamlet. The first house on the left was the Hotel Beauminet.

'Very good people,' Charlemagne affirmed as he parked the car in virgin snow behind the hotel. 'I know them both well. Paul Vennier is one of the best, and so is Madame Vennier, even if she is Swiss.' (There was a feeling in France during the war that Swiss people might be pro-German.)[1] 'The pair of them will tell you all you want to know, you'll see.'

He was perfectly right. The Venniers were only too glad to talk. They knew the number of Luftwaffe men stationed in the big square of farm buildings at Le Presbytère (or Theuville), but they had not heard of the lone villa being occupied. Of course it was a military area and no civilian had been up there for a long time. At Le Presbytère, yes, with food and other supplies, but never at the villa, nor in the radio station near the cliff's edge. Then there was the guard post in the villa Stella Maris, down by the beach. It and the machine gun posts were not, they thought, completely manned all the time. But they certainly could be manned at a few minutes notice, and there was always a guard mounted there, about ten soldiers. The Bruneval garrison was an infantry platoon under a sergeant, an efficient and energetic soldier; they were all quartered, to the Venniers' shame, in the hotel. Yes, the troops seemed good. Neither young nor old, and they were kept on their toes, since the whole countryside swarmed with German units, some of them armoured. At the end of a long and amicable conversation Pol turned to Charlemagne.

'Let's go and look at the sea.'

'Verboten!' Vennier said. 'The beach is mined.'

'Let's have a look anyway,' Pol insisted.

An icy wind tore up the gorge at them as they walked downhill with rather grim houses, Northern French gable-end romantic, their doors and windows sadly in need of paint, on either hand. Just before the barbed-wire entanglement across the road the cliff path climbed almost vertically on their right towards the radio places. Many healthy looking rabbits were sitting on the edges of their warrens. On Pol's left now was the seaside villa, peeling cement, Stella Maris. A tall German sentry emerged from the doorway and looked sadly at them. Charlemagne was very friendly.

'Good morning Fritz,' he said in German. 'Just taking a stroll with my cousin here. He's from Paris, you know. Feels he must see the sea before he goes on home. Shut up in a dark office all day long you see. You know how they get, desperate?' The sentry was smiling. Charlemagne positively seemed to be mellowing him. 'Lucky you're here,' he now continued. 'Without you we wouldn't have dared go any farther. We've heard there are mines. Imagine that!'

'Ja, Tellerminen.'

'Nix gut. I wonder if I dare suggest such a thing . . . Would you accompany us down the shingle, just for a second. It would give this dear fellow so much innocent pleasure, I assure you.'

'Jawohl.' He was a pleasant countryman, and bored to tears in that lonely place. He pulled aside a 'knife rest' to make an opening in the barbed wire, and the pair went through. He shut it behind them and made them follow closely on the path through the supposed minefield. Soon they were on a short length of beach overhung by cliffs. It was a steep-to beach of round pebbles bigger than marrons-glacés, no comforting sand. Even on the calmest day it seemed likely that a swell would break there. Charlemagne knew it from peacetime, and confirmed that there was an undertow. An uncomfortable place for bathing. Underwater obstacles? The tide was low, and there was no sign of anything like that . . . While Charlemagne gave their German friend a cigarette, Pol stood apparently dreaming. He allowed himself the luxury of thinking that the English shore was only a hundred and fifty kilometres across the sulky-looking water. Then he turned to look up at the machine-gun emplacements, one to the south of the Stella Maris villa and not far above its roofline, and another similar post but higher up, on the north side. Nasty! He could actually see the snout of a machine gun at the nearer place, and it would have nearly an all-round field of fire. Pol shivered. He saw one soldier up there, wearing a forage cap, not a steel helmet, and several greatcoats by the look of him. A boring station. No wonder the Schloks were often hitting it up at night in the Beauminet. What else? Barbed wire? None in view beyond the rather dense barrage across the road to the beach.

On the way out Pol noticed that the obliging German crossed the supposed minefield with no sign of caution or worry. The 'mines' were there to discourage the curious, and the enemy thought the place impregnable without them. He was of a like opinion.

They walked briskly to the hotel, the wind now in their backs, and there Charlemagne found some of his regular customers. Garagistes were in great demand. The gas- or charcoal-driven cars of ordinary people needed constant tinkering; and tyres were scarcer than gold, since nearly all of them went to the occupying forces or to Germany itself. After a glass of 'calva' and thirty minutes of business talk which was excellent cover for their visit he drove his 'cousin' to the Hotel des Vieux Plats in the agricultural village of Gonneville-la-Mallet, seven kilometres inland. There the couple treated themselves to a delicious luncheon at an immoral and immodest price and without official restriction of any kind. It was the sort of black market restaurant (if a particularly good one) that existed in every corner of France during the war; it was also the type most likely to be patronised by the higher-ranking and richer German officers. With the coffee (real coffee) and calvados Pol called for the visitors' book, the livre d'or. From it, when they were alone, he copied the names of all Germans, knowing that their units could be traced in the German Army lists that were in London.

While Pol was fulfilling his mission at Bruneval, Rémy his chief, was not far away, hidden in the small café of The Guardian Angel, Marcel Legardien and his wife Suzanne. Rémy and a companion were expecting to be picked up by a Lysander of the Tempsford Squadrons that brought supplies to the Resistance.[2] Bob, the wireless operator, was also there. He had slung an aerial about the Legardiens' attic. Rémy had with him a cabin trunk filled with incriminating documents destined for the BCRA in London, and a big suitcase holding all the latest military maps of France, which were also urgently wanted across the water . . . Night after night the wind blew and the snow fell. The cloud was low. The operation had to be suspended. And when he got back to Paris he found that there was a crisis.

* * *

Oh horror! 'Hilarion' reported that those accursed German warships, *Scharnhorst*, *Gneisenau*, and *Prinz Eugen*, might well be preparing to leave Brest. Hilarion (Lieutenant de Vaisseau Jean Philippon) was the CND leader in Brest, and as a spy he was superb. He belonged to a group of fifteen hundred French sailors who, with ten Navy officers and thirty engineers, still functioned in the Arsenal. Only the certainty that some of his companions would pay with their lives for the discovery of his clandestine work tempered Hilarion's daring. His information was technical, detailed, and, so far, infallible. He had become so important to CND that Rémy had, until then, refused to let him have an operator and a transmitter. The radio-detection services of the Germans in Brest and in Lorient were known to be even more deadly than elsewhere, and the streets were filled with informers and with representatives of the Gestapo, the Abwehr, the GFP, and worst of all the Milice. Down in the basins of La Ninon the *Scharnhorst* and the *Gneisenau* were hidden by huge nets. Their anti-aircraft equipment was fired by squads of German gunners imported for the purpose, while their crews slept outside the much-bombed town. Nobody in Brest got enough sleep, since the RAF bombed most nights.

Rémy hated going there. He always feared that his visit might compromise Hilarion. But this time he had to go, and to take with him 'Lenfant', André Cholet, a balding, witty, sadfaced 'radio', one of the best. They also carried a set for Hilarion, who had chosen to use his own operator, Radio-Quartermaster Arsène Gall. Gall, a Breton Hercules, was Hilarion's personal assistant in the *Service des Jardins*, for Hilarion's duties (and later he was to be an Admiral and to command the French Mediterranean Squadron) consisted of growing radishes and lettuces in the Arsenal gardens.

Rémy, Hilarion, Lenfant, and the immense Gall seemed a crowd in the small room while Lenfant carefully described to Gall the characteristics of the British underground radio service he would have to work with, and also the extreme perils of the German listening and direction-finding service. He explained to the quartermaster his allotted 'plan', which told him when England would call and listen for his reply, and on which wavelengths. He could alter the wavelengths by changing his crystals. And here, on the plan, were his call signs, and other technical details. Gall kept nodding impassively. He was not worried. Transmissions (Lenfant went on explaining) were called schedules or skeds, and his quota would be the high one of five a week; and with the way things seemed to be going in Brest, according to Hilarion, England would probably step up that number to one a day, or even more, with a lot of emergency standby. Gall nodded. Lenfant turned to explaining the code system, Rémy listening carefully. In a danger area like Brest, they both urged their companions, messages must never exceed thirty five-letter groups. The skeds must be kept short.

Lenfant warned them that sometimes a sked would go by with no contact at all, because these were, for the safety of the operator, sets of low power quite unlike the types Gall had used on cruisers and destroyers.

Meanwhile the message from Hilarion, encoded by the four of them, lay on the table beside a pad ruled out in rectangles for the anticipated five-letter groups of incoming message. The set had been earthed to the waste pipe of the basin, and its aerial was fixed across the ceiling. The time was approaching, it was very near. Quartermaster Gall settled himself easily in front of the set, the others marvelling that ordinary chairlegs could withstand his weight. At zero hour Gall began to tap out his call sign lightly, his hand completely obscuring the key. A shrill sound of morse came to them through his earphones and he said, 'They're calling me, but they don't hear me yet. . . .' He fiddled with the tuning of his set and then announced laconically, 'Well, here we go.' The index finger of his left hand, a finger as big as a banana, began to jerk from letter to letter of his message, while the other hand tapped it out. Every now and then, but unusually seldom, Lenfant noted, he would be asked for a repeat, his headphones shrilling. Then his message was sent, and he picked up a pencil and began to print letters in the rectangles on his pad. Occasionally he would query a letter. When he laid down the pencil, there was an exchange of signals, and he pulled off the headphones.

'Satisfied?' Lenfant asked him.

'They know their business. If it's always as straightforward as that. . . .'

'Get the set away and hidden, and remember, from any place as closely watched as Brest undoubtedly is by the Funkabwehr, never send twice from the same house, the same district. Never underrate the enemy's radiogoniometrie, or they'll shop you within a day or two or three, and then where shall we all be?'

'We'll do our best, the lieutenant and I,' Gall answered seriously, watching Rémy and Hilarion, who were decoding the message from London. It was brief and congratulatory. Quality of reception, as Rémy well knew, varied from one part of France to the other. He warned Hilarion and Gall once more of the efficiency of the German direction-finding experts. If Gall heard any sudden interference while he was in communication he should shut down at once, hide the set and leave the house by his alternative route. Interference often meant that the German 'central' was trying to hold the clandestine operator at his set while the direction-finding 'gonio' vans closed in on him.

When the two visitors left by Hilarion's back door Rémy was turning over in his mind the gastronomic possibilities of Brest at that time. He heard Lenfant stammering at his side.

'God help us all!'

'What then?' Rémy asked, gripping him by the arm.

'The courtyard next to Hil's house. . . . .'

'Well?'

'It's covered with aerials, German aerials . . . .'

'I'll speak to Hil about it. . . . Do you like snails? I know a little place where they aren't bad. Just let us pray that an English bomb hasn't been there before us. . . .' Rémy remembers to this day that the restaurant was there and functioning; but there were, alas, no escargots de Bourgogne, only humble petits-gris. Nevertheless they were cooked in real butter. And they comforted themselves with a fortifying glass of 'lambic', eau-de-vie de cidre.

Hilarion took to riding about on a bicycle, accompanied by Quartermaster Gall. They carried bits of transmitter in their pockets, and others strapped to their carriers, and they never transmitted twice from the same place. Hilarion thus was able to report to London the Germans' secret preparations for taking the warships to sea. Other CND agents in the northern sector reported Luftwaffe fighter squadrons and wings moving to the Channel airfields, including several of those now flying the damnably good Focke Wulf 190.

Back in Paris, Rémy had an appointment with Pol at 'Jeff's'. Jeff was Madame Lucienne Dixon, a Frenchwoman married to an American. She lived at 1, Rue General Largeau, near the Porte d'Auteuil. Pol was ready with his report on Bruneval (or Theuville). They worked on it together, trying to condense it.

'Even so, it's going to make the hell of a vacation,' Pol said. (Nobody ever used the word transmission.)

Uneasily agreeing, for the sending of any message—let alone an unusually long one—was incredibly dangerous in Paris, Rémy went off to his own flat where he checked and double-checked the coding of the two messages. He was there, sitting with Madame Edith and the children, when Bob arrived, filling the place with cigarette smoke.

'Be careful,' Rémy said, handing over the flimsy bit of paper.

'Don't worry,' Bob answered in his usual carefree manner. 'Les Boches sont trop c. . . .' And that same evening, as though it were the most natural thing in the world, as though Paris were not hotching with radio-detection units, he informed Rémy by telephone that both Bruneval messages had safely passed to London.

NOTES TO CHAPTER 10

1. The author, who worked for SOE near the Franco-Swiss frontier, sometimes on, sometimes over it, constantly met this prejudice, which was never once justified. The French, like the British, mistrust 'foreigners'.

2. For the work of the Tempsford Squadrons see Foot, *SOE in France*, and Cowburn, *No Cloak, No Dagger*. 'Over a hundred successful pick-up sorties to France were made for SOE, delivering over 250 passengers and bringing nearly 450 out, for a total loss of two Lysanders, one pilot, and two agents.' Foot, *op. cit.*, p. 88.

## 11

# Worth Matravers

*1940/41*

Worth Matravers looks over the edge of the Dorset Downs to the tidal swirl of the Channel near St Alban's Head. The village's largest house is the modest vicarage; its inn, the *Square & Compass*, is discreetly sited on the left as the lane enters the village. The ground, chalk and flint, is not rich, but it gives a swept feeling of durability, of calm; the sky is vast at Worth. The nearest town, squatting under the cliffs and invisible from the village, is the small one of Swanage.

It was to this corner of ageless England that the Radar men came to get on with their work. First they had been at Orfordness, where Radar had become England's hope, then at Bawdsey where, with the birth of the Chain Home it became her defence in depth. From Bawdsey the brilliant Watson-Watt had been promoted to London, and A. P. Rowe, a much younger scientist, became superintendent. At the outbreak of war the politicians feared that the supposedly omniscient Germans would instantly wipe out Bawdsey, killing its useful (indeed essential) scientists. At the same time the Nazis, who had little liking or respect for Germany's wealth of scientific talent (though much of it had seeped away before the war to America, England, and Sweden) were drafting young scientists into the forces in the hope that they would meet an early and ennobling death. The Bawdsey Establishment was banished to the safety of Dundee, where it was not even expected by those designated to be its hosts, and where its gear had to stand out in the mist and the rain protected only by packing cases. The climate was unsuited to the flying side of their work. And it was part of the wartime scientist's make-up that he worked best in the thick of things. Wisely, the Air Ministry listened to Rowe's complaints about the immolation in Scotland, and on May 5, 1940, the Establishment moved south in convoy to settle in and round Worth Matravers. Six months later it was given the three-letter name, TRE (Telecommunications

Research Establishment), that became known and admired, not by the general public who heard nothing of it, but by those in England and, later, America who were running the war.

At Worth Matravers the scientists, whose numbers kept increasing, worked in huts protected by earth banks against bomb-blast. Their presence on the headland when the Battle of Britain came, quite soon after their arrival, was advertised by two Radar towers, one of 350 feet and the other of 240 feet. They found that they had too many raid warnings and other interruptions, so they dispersed locally into an empty private school, Leeson House, between Worth and Swanage, then into another school, Durnford House. Leeson House stables, which they used as a laboratory, overlooked Swanage and Swanage Bay, with a view away eastwards to the Needles and the entry to the Solent. Still the expansion continued. Rowe, a Government servant, found himself sitting on a hillock that heaved and pulsed and grew. The Establishment overflowed down the hill, moved into Swanage, and began to recruit assistants there. Rowe learned that the scientists were perfectly willing to work in the most unlikely and unsuitable corners and at all hours, provided they were given enormous supplies of electric power. At the start there were no holidays, no weekends, no days off. This worried Rowe. Left alone, those people would work until they became jaded and stale. He decreed that Saturdays would be holidays.

The Sunday Soviets, which played a significant part in the extraordinary development of TRE, had begun at Bawdsey Manor. Bawdsey, with its cricket pitch, its beautiful grounds, the proximity of the sea, and the variety of its war-time staff, most of them young, had a university-like atmosphere. Then it was near Cambridge University which in the Cavendish Laboratory held the remarkable pool of scientific talent created there by Rutherford. There was much mixing between the two centres, and possibly that was why the Sunday Soviets got away to such a good start. Now they were held in Swanage, and their importance became increasingly recognised. Senior officers, members of the Cabinet, scientists, jaded by the week's worries in London, would get themselves down to a Swanage hotel by Saturday evening and on Sunday morning would be there in Rowe's office, where *anything* could be said. The senior officers could speak of their needs in the way of equipment while junior ones, many of them straight out of action with Fighter or Bomber or Coastal Command, could tell of faults and ask for improvements. As well as learning from their guests, the scientists of TRE were able at the Soviets to explain and to 'sell' their newest ideas. The Soviets were an immense time-saver. There was nothing remotely like them in any other service scientific establishment in England or in any other country.[1]

When Rowe became superintendent at Bawdsey in 1938 there were a few draughtsmen in a small drawing office and some twenty skilled mechanics to make experimental equipment. By the end of the war Rowe was superintendent of a TRE that employed two hundred draughtsmen and that had five hundred men of different skills in its model shop. Swanage had soon failed to accommodate TRE's new engineering unit, which was built at West Howe, near Bournemouth. As well as inventing, testing, and improving, the Establishment understood that its task was to shorten the period between the birth of an idea and its full-scale use in operations. Consumer reaction was studied and was moulded. Air crews were informed of devices in the pipeline and were taught how to use them with electronic simulators made by TRE. The Establishment made its own instructional films. Its small Radar school set up in Swanage grew until thousands were passing through it. Rowe remembers 'a front row of Air Vice-Marshals and Air Commodores, sitting at the feet of flannel-bagged lecturers. . . . This could not have happened in Germany'.[2]

Apart from helping technically in the Battle of the Beams, TRE's first pressing task at Worth Matravers was to produce an efficient means for the night fighter to get within range of the night bomber.

Airborne Radar, in which Britain had a lead over Germany, had been initiated by E. G. ('Taffy') Bowen in the autumn of 1936 at Bawdsey. Bowen's work was supervised by Robert Watson-Watt at the start, and Watson-Watt said that airborne Radar must be designed to throw out a long, narrow beam, but to have such a Radar much shorter wavelengths were wanted. Bowen's first Air Interception (AI) set worked on a 6·8 metre wavelength, but he soon had one working on 1·5 metres. Tizard and Dowding, both realists, knew that in concentrating on Britain's day defences they had left their country open to enemy night attacks. There just had not been time to do more. They felt that if they had not concentrated on being able to beat off daylight air attack, the night attacks might never have come. From the outset they gave Bowen every help. In 1938 a special RAF flight was attached to the Bawdsey unit to work with him. The flight had gone north with the scientists and now, its numbers increased, it had settled at Christchurch. Soon it got too big for that aerodrome and moved to Hurn (now Bournemouth Airport).

By July 1940 Bowen had an AI set, the Mark III, working in a Blenheim. The Blenheim was too slow for a night fighter, and the Beaufighter, which was to be its replacement, was giving a lot of development trouble. As to the 1·5-metre AI set, its maximum range was two miles, which was on the short side, and its minimum range was eight hundred feet, which was on the long side. Also the range of the bomber from the fighter had to be less than the altitude of the two aircraft or the unwanted return signals from the earth were powerful enough to obliterate the return signals from the bomber. In other words, if the hostile aircraft came in at two thousand feet, the night fighter's Radar range was reduced to less than two thousand feet.

It was understood that the 1·5-metre AI sets could never be entirely satisfactory. They were vulnerable to enemy jamming, the beam was far too wide for accuracy, and much of its energy spilled to the ground. This could have been improved by having aerial clusters on the noses of the fighters (such as the Germans fitted later in the war on the noses of their Ju 88s). But such aerials had a sorry effect on the speed and handling of the aircraft. No, what was needed in those small airborne sets was a *very* short wavelength, ideally one of ten centimetres. To get that a valve of small size which produced fantastic power would have to be devised.

And devised it was, at Birmingham University by John Randall and Henry Boot.

A deal of the credit for achieving the wonder valve must go to the Admiralty, which, before the advent of Radar, was the most valve-conscious service department. In 1938 Dr C. S. Wright, its Director of Scientific Research, had predicted correctly that the side that developed valve power on the shortest wavelength would win the war. After a sticky start in Radar, the Royal Navy had two ships, *Rodney* and *Sheffield* fitted with experimental early-warning sets by the end of 1938. And that year the energetic man who had built up the Admiralty's scientific power in the Thirties, Sir Frederick Brundrett, more or less collared, and therefore unified, valve production. In the face of opposition from the Air Ministry, Brundrett organised the setting up of a single inter-service organisation, which he ran himself (with a committee). At the outset of war he called together the major valve manufacturers, whose production was hampered by industrial and scientific security. Brundrett persuaded them to lose their inhibitions about exchanging secrets—for the good of their country.

Having got round the manufacturers, Brundrett called for major research. He got teams going at the Cavendish, the Clarendon, and at Birmingham and Bristol Universities. At Birmingham an Australian, Professor M. L. E. Oliphant, headed an impressive group of young men, most of them nuclear physicists. Oliphant and some of the others set to work to develop a valve called the klystron which had been recently produced by the brothers Varian at Stanford University, California. The Greek word klystron means, charmingly, incoming waves on a beach, and in the valve electrons were driven along a passage past or through resonators. Two of the Birmingham physicists, Randall and Boot, neither of whom was engaged with Oliphant on the klystron, had the notion that the 'waves on a beach' electron principle might be applied to another instrument, the magnetron, originated in America in 1921.[3] They designed a valve, the cavity magnetron, and they built it in proper Heath Robinson fashion. Air was drawn from their experimental model by a continuously working pump. The ends of the valve were closed by embedding halfpennies in sealing wax. A clumsy laboratory electro-magnet was used to provide the magnetic field.

They were ready to test the valve on February 21, 1940, and the cavity magnetron led off at a hell of a gallop. Two car headlights connected to it flared and burst. Two bigger headlights could not take the power. Low pressure neon lamps were connected, and they showed that the halfpenny-and-sealing-wax contraption was producing four hundred watts on a wavelength of nine centimetres. Brundrett at once reinforced success with support from the Admiralty and from the General Electric Company at Wembley and the British Thomson Houston Company at Rugby, as well as from the other universities. When GEC Research Laboratories under C. C. Paterson developed the first production model it produced ten *thousand* watts.

TRE experiments with this new marvel began in July 1940 under P. I. Dee, another pre-war nuclear physicist. One of his helpers was Bernard Lovell. The ledge by Leeson House stables with its view away eastwards was an ideal ground site for experiments with centimetric Radar. 'Soon,' Rowe says,[4] 'there was a line of trailers there, looking like caravans, each with a metal parabolic mirror overlooking the town

[Swanage] and the sea. This became known as Centimetre Alley.' An experimental AI set using the magnetron detected a 'hostile' aircraft at a range of six miles in August 1940. But it was not until March 1941 that a prototype centimetric AI set went up in a night fighter. Through the long earlier stages of the night battle the pilots had to make do with blunter tools.

It may seem odd that if Radar could work in daylight it did not readily work in the dark. But during the day Chain Home and its ancillaries could put fighters on to their targets with height errors of a couple of thousand feet, distance errors of a couple of miles; the fighters scrambled and the human eye added to the human brain took up the slack. Without vision, man is one of the more helpless animals. Then there was an extra problem in that the Germans (like the RAF in the early stages of their attack on Germany) favoured dispersed night-bombing attacks. TRE calculated that during an average night raid over England in 1940/41 there was one German bomber to every nine hundred cubic miles of darkness.

There were two sides to catching them. The pilot of the night fighter had to 'see' the enemy with his AI set. But before that he had to be directed to within AI range by controllers on the ground. Here Rowe explains:[5]

'The solution was reached with the Plan Position Indicator, which was installed in a new form of Radar set known as the GCI (Ground Controlled Interception). The controller had before him the screen of a cathode-ray tube similar in size to that of a domestic television. Marked permanently on this screen was a map of the surrounding area up to distances from him of about fifty miles. Positions of the hostile bomber and the friendly fighter were shown by bright spots on the tube face which, in addition to the map, was marked in grid squares. By calculating the speed and direction of flight of the enemy, the fighter could be directed by the shortest route. The IFF (Identification Friend from Foe) set carried in the fighter caused the friendly spot to give a characteristic signal every few seconds.' (IFF was developed at Bawdsey in the very early days and was subsequently improved. It was carried by all RAF aircraft. There arose among British bomber crews a superstition that if they kept IFF switched on over Germany and Occupied Europe the set interfered with the ice-blue radar-controlled German 'master' searchlights, which probed their blue fingers vertically into the sky and then, with demoniac speed, fixed on an aircraft. If you were located by one of these, a team of ordinary searchlights caught you in a cone, and escape from the consequent flak was horribly difficult. Keeping IFF on over Germany—although it had been devised for the return to England—became a habit that cost British lives. German night fighters were fitted with a device, later in the radio war, that could home on IFF.)

The first ground-controlled interception (GCI) set was built at Worth Matravers and began its series of tests while TRE had agreed to put through a 'crash programme' and produce twelve GCIs before the end of 1940.

'During that lovely summer,' writes A. P. Rowe, 'we often saw hostile day bombers passing over us. One day we counted seventy-two of them. . . . There were times when, during paper-bag lunches on the cliffs by St Alban's Head I saw aircraft with smoking tails take their last dives into the sea. But these events were little concern of ours, for our eyes were on the coming night-bombing war. We had a GCI set from which fighters could be controlled; we had the co-operation of AI-fitted night fighters at Middle Wallop; and we had enemy night bombers. After our day's work many of us went to our GCI station, hoping that this would be the night on which civilian scientists working as controllers at a research station would be the means of bringing down a night bomber. . . . Bomber and fighter would be tracked and their heights assessed. When the fighter had been put on the tail of the bomber the GCI station would give the signal, "Flash Weapon!", and the fighter crew would switch on the AI set. . . . It is sad to record that those who had evolved AI failed to obtain a kill. They were replaced by RAF controllers who, in fact, did the job much better.'

On October 16, 1940, the first CGI set was delivered by TRE for service and the last of the twelve sets was delivered six days after the New Year. The sets, Rowe says, 'were often hand-made with odd bits and pieces. . . . There was still, that winter, a period of frustration and training before results came. Then enemy losses increased month by month until in May, 1941, 102 night bombers were shot down by fighters and 172 were assessed as probably destroyed or damaged. During this period the casualty rate suffered by night bombers rose from less than half of 1 per cent to more than 7 per cent. . . .'

* * *

Bomber Command's three navigational aids of the war (the Command began it with none) were devised and developed by TRE at Worth Matravers. The first, Gee, a system of navigational pulses, was soon jammed by the enemy but remained invaluable as a navigational guide. (Someone in the Navy suggested that D-Day should have been called Gee-Day—the ships used it too.) The second system, called Oboe was the child of Alec H. Reeves, an inventor in the pulse techniques field, who joined TRE at Swanage from Standard Telephone. If Reeves had not had the enthusiastic support of a distinguished operational pilot—our friend H. E. Bufton—Oboe might never have been used, since it and Reeves roused fierce opposition. For example a 'high-ranking official' in the Ministry of Aircraft Production wrote, 'I regret having to do this, but I am sure it is time to say quite bluntly that these disquisitions from TRE on Oboe are becoming ridiculous. If they came as inventions from the outside public, and not from official sources, they would be rejected without hesitation. . . . If I had the power I would discover the man responsible for this latest Oboe effort and sack him, so that he could no longer waste not only his time and effort, but ours also, by his vain imaginings.' The letter was passed to Rowe by its recipient, a 'higher-ranking official' in the same Ministry—and Rowe drafted more staff to support Reeves. Oboe, which worked with two ground Radar stations, was the most accurate navigational aid of the war, though its range was very limited—a maximum of 270 miles. It was Oboe, carried in another remarkable invention, the Mosquito, that brought destruction to the Ruhr. As for $H_2S$, the third system, it came from the Leeson House stables above Swanage. For when Dee and Lovell tried their experimental centimetric AI sets there they registered ships entering and leaving the Solent. Dee took a set up in one of his Blenheims and had the aerial looking slightly downwards instead of ahead. The set's revolving arm blocked in a Radar map, and on the display appeared—Southampton. This meant that with $H_2S$ a bomber could carry a self-contained navigational aid, one that could reveal to the aircrew an actual map of their target. There was one snag, and it was a horror. The cavity magnetron was a solid block of copper in which the highly secret cavities had been hollowed out; it appeared to be indestructible. Farnborough tried in vain, and with live experiments, to destroy it. In the course of one such experiment a ten-foot hole was blown in the fuselage of a captured Junkers 88. When the experts examined the remains of the magnetron on the ground they saw that it was still possible to understand how it worked.[6] Surely it was unthinkable that this incredible device should be virtually handed over to the Germans?

Yet another operational question had to be answered. Should not centimetric Radar be reserved for the U-boat war? When Bowen was developing AI in 1937 the early set's weakness, the return of strong echoes from the ground, showed him its possibilities over the sea. The marine version of the set was called ASV (Air to Surface Vessel). By the summer of 1940 Coastal Command had Mark II ASV which was technically capable of locating a surfaced submarine at four miles range, and homing on to it. At that time the Germans had comparatively few submarines. However the U-boat fleet was quickly increased, and its main operating bases were on the French Biscay Coast. The U-boats had to surface to run their diesels and charge their batteries. If they surfaced in daylight the British aircraft could, and did, catch them in the Bay, using ASV. The first German answer was to surface at night. The British replied with the airborne Leigh Light which was bright enough for the aircraft, having located the submarine by ASV, to put in an attack. But all that the Germans had to do, the British realised from the start, was to put in each U-boat a simple radio-listening device which would receive energy from the wide beam of the 1·5-metre ASV. This could give visual or aural warning, or both, when an ASV-equipped aircraft approached, and the submarine could dive to safety. (The U-boats were so noisy when surfaced that the crews did not hear approaching aircraft.) But if the cavity magnetron could be kept exclusively for Coastal Command there was no reason why it should fall into German hands, and the U-boat war would be won, since the Germans' radio device, effective against ASV, would be useless against $H_2S$/ASV.[7]

It seemed clear to the majority of scientists and war leaders who knew the problem—the indestructability of the cavity magnetron—that Coastal Command could make more valuable use of $H_2S$ than Bomber Command. But Lord Cherwell (Lindemann had now been raised to the peerage), with what Jones has described[8] as 'his emotional commitment to make bombing more scientific, combined with his inclination to attack rather than defence', insisted that the bombers, as well as the anti-submarine patrols, have it. He won the day and one of the first bombers fitted with it, a Stirling of No. 7 Squadron, was shot down near Rotterdam. The aircraft was an eighty per cent wreck.

Two of the crew had survived but 'both have obstinately and consistently refused to make any kind of statement', Engineer-Colonel Schwenke stated in his report. Plendl, Lorenz, Martini were called to the enquiry into *Rotterdam*, the German name for $H_2S$. Telefunken were ordered to make six Radar sets of the same centimetric type, and two additional sets, the first, Naxos, a detector, and the second, Korfu, a direction-finding receiver to be fitted in night fighters. When Göring read the initial report on *Rotterdam* he said: 'We must admit that in this sphere the British and Americans are far ahead of us. I expected them to be advanced, but I never thought that they would get so far ahead. I did hope that even if we were behind we could at least be in the same race.' The first *Rotterdam* salvaged by the Germans was soon completely lost in a bombing attack on Berlin in which the Telefunken works was badly mauled, but that same night the Germans obtained a second $H_2S$ set when one of No. 35 Squadron's Halifaxes was shot down over Holland. Careful interrogation of RAF prisoners, too, soon gave results.

There is a story that Cherwell invented the name $H_2S$, but this is incorrect. . . . Dr R. V. Jones found that TRE were working on a new aid which they called 'TF'. Jones correctly guessed that this stood for 'Town Finding' and that it related to Dee's vastly successful experiments with centimetric Radar. He pointed out to Cherwell the insecurity that arose from TRE's frequent practice of naming devices by the initials of words that described their purposes. Cherwell acknowledged this, and said he would get it altered when he got down to Swanage (he was going to a Sunday Soviet). When he returned he said that he had been successful, and that the new name was $H_2S$. He enquired if Jones, that remarkably perspicacious Jones, could guess the connection? Jones could not.[9]

'It's very clever,' Cherwell said. 'It stands for Home Sweet Home.'

When Jones next visited TRE the change of code-name cropped up in conversation, and he found that the new name's derivation was a scientific one. When Cherwell had asked them to change 'TF' they had considered it over lunch, and one of them had recalled an incident of a year or so earlier. . . . Cherwell had been shown the possibilities of a certain device, but had not seemed interested. On a subsequent visit, after being shown everything else, he asked what had happened to that particular device. In the meantime, relatively little had been done about it, precisely because Cherwell had shown no interest. They did not think it polite to tell him that, in so many words, and they tried to give other reasons for the lack of development. Cherwell became extremely annoyed, and said he was thoroughly upset by such prevarications in a scientific research establishment in time of war. '*It stinks!*' he exclaimed. '*It positively definitely stinks!*' The TRE executives at lunch therefore enthusiastically received a suggestion from one of their number that 'TF' be renamed '$H_2S$'. What they did not reckon with was that Cherwell should at once ask, 'But why $H_2S$? What is the connection between $H_2S$ and this device?' There was an awkward silence until one of those present imaginatively suggested 'Home Sweet Home'.

## NOTES TO CHAPTER II

1. 'I will never forget the invigorating atmosphere that prevailed at Bawdsey and, later, Worth Matravers. Here was brilliant individualism harnessed to make a great team without loss of individual freedom and initiative. This freedom of individual thought was given its full expression in those stimulating weekly conferences which, with the paradoxical humour which is so typical of our people, were called Sunday Soviets.' Excerpt from the Foreword by Lord Tedder, Marshal of the RAF, to Rowe's *One Story of Radar*. . . . Paradoxical indeed; according to this writer's observations scientists, if they descend to politics at all, incline to Conservatism rather than Socialism.
2. Rowe, *One Story of Radar*, p. 95.
3. Development of the cavity magnetron: for an excellent non-technical description see Clark's *Rise of the Boffins*, pp. 128–36.
4. Rowe, *op. cit.*, p. 82.
5. Rowe, *op. cit.*, p. 64.
6. Price, *Instruments of Darkness*, p. 123.
7. In August 1943, Hitler admitted a 'temporary setback' (Allied shipping losses had fallen from 400,000 to 40,000 shipping tons per month) caused by 'a single technical invention of the enemy'. This was $H_2S$/ASV.
8. Jones, *Minerva*, Vol. X, No. 3, p. 450.
9. Jones, letter to the author.

## 12

# Sidney Cotton

*1939*

Alfred J. Miranda, an American business man, on September 14, 1938, cabled F. Sidney Cotton, an Australian business man with offices in St James's Square. Miranda said he would arrive in London the following day, and he urgently wanted Cotton to accompany him to Paris. Miranda dealt in many things, including aeroplanes and guns, whereas Cotton's business at that time was in a new type of film, Dufaycolor. In Paris Cotton met Paul Koster, Miranda's agent, 'a man of seventy, but spry and animated'.[1] In the course of two meetings Koster sounded out Cotton's reactions to the rising German menace. He received Cotton's assurances that he would be ready to take part *immediately* in the inevitable war with Germany. Koster told him he would be contacted on his return to London.

The telephone rang. The speaker said he was 'a friend of Paul', and asked permission to come round right away. One hundred and twenty seconds later Cotton's secretary brought in the visiting card of Major F. W. Winterbotham. 'I looked up to see a man of about my own age,' (then forty-four) Cotton says,[2] 'dressed in a grey suit which toned with his grey eyes and greying hair. There was a look of determined discretion about him.'

Winterbotham told Cotton that he represented official intelligence organisations in England and in France. Now that Germany was expanding in all military ways and at the same time had clamped down the strictest peacetime security yet seen anywhere, what was needed was a privately owned aircraft able to take clandestine aerial photographs of German and Italian fortifications, aerodromes, and factories. Cotton was well known as a flying man on the Continent, and he also had bona-fide business interests there. Would he be willing to let Winterbotham buy him a suitable aircraft, and if so, what would be the best for the job?

'A Lockheed 12A,' Cotton said. He picked up a telephone and asked to be put through to Miranda in New York. Miranda said he would buy him a Lockheed at once.

What remuneration would Cotton require for such work, Winterbotham asked without fuss or embarrassment. None, Cotton answered, so long as his out-of-pocket expenses were met. In what immediately followed they did not see eye to eye. Winterbotham explained that Cotton need only be the official owner of the spy aircraft, because the Intelligence flights were to be made under the aegis of experts, the French Deuxième Bureau, who had 'a predominant interest' in the flights.

In January, 1939, the new Lockheed arrived. Cotton at once flew it solo, and was delighted with it. He was more reluctant than ever to hand it over to the French, who produced ham-fisted pilots and photographers whose methods and equipment caused Cotton's hackles to rise—a posture they were never laggard in assuming. There was a nightmare flight over the Rhine with a Deuxième Bureau camerman astern in the cabin. The Frenchman had established contact with Cotton in the pilot's seat by attaching long strings to each of the Australian's elbows. Responding to the tugs and wondering where the German fighters were, Cotton wove a zig-zag course across the map. He demanded to see the resulting photographs, refused to be denied, and was dissatisfied with what he saw. He proposed to the French that he should take over the photographic side, which he understood as well as anybody in Europe, including Germany. Instead of their one slow and massive camera he would make up a frame to take three of the RAF's F.24 cameras, two angled outwards and the third looking straight down, and covering several square miles at each exposure. He knew that the RAF results with the F.24 were disappointing, but he also knew he could improve on them dramatically by using Leica film and fine-grain developer, and by playing warm air round the cameras and thus cutting out condensation. The French refused to do things his way. So he told them to take the Lockheed; and he told Winterbotham that he was telephoning Miranda to buy another Lockheed. When it arrived he would really get some work done. There was not much time. Winterbotham agreed.

The second Lockheed, even better than the first, reached Southampton early that May and Cotton fixed it with his remarkable blend of ingenuity and thoroughness. Extra fuel tanks were fitted and 'teardrop' windows so that the pilot could see below and astern (one of Cotton's own patents). Airwork, a private firm, at Heston made him 'secret emplacements' for five cameras. Metal slides, almost impossible to detect, covered the embrasures when the cameras were not in use. There were three F.24s in the belly, and a Leica mounted in each wing. All controls for the cameras were electric, and were connected to the pilot's seat. The cameras and films and even the portable oxygen equipment could be packed away in suitcases which were covered with travel labels to make them look like innocent luggage. Lastly, colour: one day at Heston he watched the Maharajah of Jodhpur's private aeroplane taking off, and when he looked up a few seconds later he could not see it. Cotton deduced that the Maharajah's colour, a pale duck-egg green, was wonderful camouflage. He had the Lockheed painted just a shade paler than that, and patented the colour under the name 'Camotint'.

Winterbotham's first assignment for the gorgeous new spyplane was a long one, covering the Italian military scene outside Italy proper. Cotton chose Bob Niven, a young Canadian, for his co-pilot. They were refused (with full tanks) a Certificate of Airworthiness. But Cotton, a hard one to baulk, decided that 'I could hardly be blamed if some of the tanks were filled in error.' All were filled, and he and Niven took off with plenty of lift, and room to spare. On June 14 in Malta Cotton recorded: 'Introduced today to an RAF pilot named Shorty Longbottom—height five foot four—a fine young pilot with a slide-rule mind, keenly interested in my ideas on aerial photography.'[3] Shorty could only get leave to do a brief flight with Cotton and Niven over Sicily. They photographed Comiso, Augusta, Catania, and Syracuse, with excellent results. 'Very glad,' Cotton wrote, 'to have Shorty with me, as he checked over the working of the cameras and gave me many useful tips on their operation.' Cotton flew on for eleven days, staying only in the best hotels, and making photographic cover of the Eastern Mediterranean, North Africa, and Ethiopia. It was time to return to England and go to work on Germany.

Most conveniently, the Germans were genuinely interested in his colour film. Dufaycolor's agent in Berlin had flown in the Richthofen Circus with Göring, and knew all the senior Nazis. When Cotton first arrived at Tempelhof Aerodrome, on July 26, 1939, there was a jack-booted guard of honour. And for a short period in which he took many clandestine photographs, Cotton and his 'kolossal' Lockheed were frequently seen in and over the Reich. Both were nearly casualties at the beginning of the war. Cotton had had a plan to bring Göring to England to convince him that if Germany attacked Poland the English would fight to the death. Göring had first agreed to a swift journey over in the kolossal Lockheed, which had, in German eyes, assumed an almost ambassadorial status. Chamberlain and Lord Halifax had agreed to such a meeting, and had made plans for a reception at Chequers that Cotton considered to be far too formal. Winterbotham hated the notion of the trip. He believed that the Germans would expedite their invasion of Poland, and if this became certain he said he would send Cotton a warning telegram:

MOTHER IS ILL
MARY

That telegram was delivered to Cotton at the Adlon Hotel, Berlin, on August 24. The following day, when the secret order to begin the attack was issued and then withdrawn (because of Mussolini's disapproving intervention and of the signing of the Anglo-Polish pact) a further telegram arrived:

MOTHER VERY LOW AND ASKING FOR YOU
MARY

In the end, after interminable delays, Cotton and Niven were allowed to take off in the Lockheed. They had to follow a narrow prescribed route out of Germany, but while over the Dutch border, East of Groningen, Cotton saw the German battle fleet anchored in the Schilling Roads outside Wilhelmshaven, and photographed it.

* * *

The above will have given some idea of the enthusiasm with which Sidney Cotton tackled any enterprise. He was a big man (he weighed a hard fifteen stone) with the needling self-assertiveness of a small man. The strongest bent in his character, though, was one of absolute independence. He was a loyal friend, one who preferred giving to receiving. He was an affectionate child; yet when he insisted on leaving his father's cattle station and fruit farm in Queensland for a life with aeroplanes and adventure he spurned his more-than-generous £1,000 a year paternal allowance. Characteristically, after a good war as a pilot in the Royal Naval Air Service, he quarrelled with his commanding officer over what he claimed to be a bad decision by the latter, and resigned his commission.

By far the most typical aspect of Cotton's First World War experiences was his invention of the Sidcot suit. One cold day in 1916 when he was working on the engine of his fighter, a Sopwith one-and-a-half-strutter, there was an enemy alert. He took off as he was, in greasy overalls. When he got back to the Mess he found that he was quite warm whereas his brother pilots in full flying kit, had nearly frozen solid in their open cockpits. That made him think. He got leave, and asked Robinson and Cleaver in London to make him a flying suit of his own design with a fur lining, a layer of airproof silk, and an outside layer of Burberry material, the whole made in one piece. The neck and cuffs had fur inside 'to prevent the warm air from escaping. I had deep pockets fitted just below either knee so that pilots could reach down into them easily when sitting in cockpits. I asked Robinson & Cleaver to register my design' (Cotton was then aged twenty-two) 'and for a name I took the first three letters of each of my names. . . . My father had drilled it into his family that none of us should ever try to make money out of our country's need in war, so I never made a penny out of the Sidcot suit, nor did I make any kind of claim after the war.'[4] This despite the enormous success of his suit, which was adopted by the Royal Flying Corps. Baron Richthofen, the greatest fighter pilot of that war, was wearing a Sidcot when he was at last shot down. It was the first flying suit to cross the Atlantic, on Alcock and on Brown, and there was hardly a man who flew for the RAF in the Second World War who did not wear one at some time.

Between the wars Sidney Cotton mixed frequently successful business activities (for a period he sold American inventions to the British and British inventions to the Americans) with flying adventures. When still a young man, using capital he raised personally, he flew the Newfoundland mails, organised an aerial survey of the island, and acted as spotter for the sealing fleet. His description of his work there, reads like an excerpt from a novel by the late Nevil Shute. Cotton learned aircraft and flying the tough way. He truly loved them, and he was an expert. While he was devising means of making his aircraft easier to start when the thermometer registered thirty below, he might also be thinking about a venture on the London Stock Exchange. For his aerial survey of Newfoundland he got backing from Lord Northcliffe, from Sifton Praed of St James's, the map people, and from a Welsh mining syndicate. Then, quite suddenly in 1923, Cotton sold out his stake in Newfoundland, where he had five operational aircraft, a sizeable working yacht, shore establishments, and a big timber business. The decision to sell was made when, flying an Avro on a seal-spotting trip, two hundred miles from land, the engine began to falter. 'No one knew what route I was on and I had no radio. I'd been taking this sort of risk for three years and so far I'd been lucky, but now the prospect of freezing to death brought me to my senses.' There speaks the born gambler. He managed to nurse the ailing Avro back to his base at Botwood, and left Newfoundland.

* * *

When war came Cotton's association with Winterbotham continued. 'The RAF are having camera trouble,' Winterbotham said one day. 'The First Sea Lord wants some photographs of foreign ports, the RAF can't get them, and everybody is raising merry hell.' He asked Cotton to go next morning to talk with the Director-General of Operations, Air Vice-Marshal Richard Peck.

While Peck was extremely good mannered and diplomatic, Cotton sensed hostility in the atmosphere; and hostility was there. Bad feeling existed between the Admiralty and the Air Staff on the subject of photo-reconnaissance. The bluff sailors were apt to contrast at inter-service meetings in Whitehall the consistent early failures of the RAF in that field and the almost nonchalant ease with which 'that fellow Cotton' seemed to get excellent pictures—a civilian in a civilian machine! Peck now asked Cotton if he had any special equipment. No, Cotton answered. Then how did he explain the excellence of his pictures when the cameras in the Blenheims were freezing up all the time? Cotton explained that it was not the cameras that froze, but the condensation round them.

A further meeting was arranged, and there Cotton met an even higher-ranking officer, the Vice-Chief of the Air Staff, Air Marshal Sir Richard Peirse. After many introductions to blank-faced officers in blue, Peirse told Cotton that aerial cover was urgently needed of Flushing and Ymuiden. RAF crews had tried repeatedly to get this cover and had failed. Could Cotton make any suggestions?

'Lend me a Blenheim and I'll get you the pictures right away.'

What a thing to say! How could a civilian fly around in a military plane? If he were shot down and taken he would be executed as a spy. Cotton spoke up. He wouldn't mind taking such a risk. He'd been taking it for some time before war was declared, as the Air Marshal was aware. Presumably they had sent for him as the photographic specialist that he was, just as, if he wanted a pair of boots he went to Maxwell's in Dover Street. Couldn't they let him get on with the job of taking the pictures. That would save a lot more talk, and it would also show whether he could do it or not. But all that the meeting decided was to have a further meeting next morning with some of the Blenheim pilots present, and an expert from the Royal Aircraft Establishment, Farnborough, who, Cotton suspected, was being called to 'disprove' his condensation theories.

Depressed by all that waffle while pictures were urgently needed, Cotton stood in his office looking out over St James's Square. 'The morning mist had cleared. It was a lovely warm day.[5] I watched the fleecy clouds as I pondered a means of breaking the deadlock. . . . It was one of those days when anyone who loves flying longs to get airborne. . . . Why not? I rang Bob Niven at Heston and asked him to get me a flash weather report for Holland. The weather there was much the same as London, and the big woolpacks of cloud would give us the cover we needed. I told Bob to get the Lockheed out and warm her up ready for take-off, and as neither of us had had lunch, I asked him to get the airport restaurant to pack us a hamper. I rang Winterbotham and asked him if he knew of any German fighter patrols along the Dutch coast. He said it was quite possible if there were German naval movements. We had no photographic processing facilities at Heston, so I asked Winterbotham to arrange for Farnborough to develop and print some pictures later in the day, warning them it was a rush job, and they might have to work through the night.' Cotton then told his secretary to inform any callers that he would be unobtainable until the following morning. Kelson, his chauffeur and personal servant, drove him to Heston. The streets and roads were empty, and the Lockheed was only just warmed up when they arrived. Cotton and Niven took off less than an hour after the ending of the unsatisfactory, indeed hostile, meeting in Air Vice-Marshal Peck's office.

Immediately after take-off Cotton asked Heston Control to tell Fighter Command that White Flight, his normal protective code for the Lockheed, was going out to sea off the Kent coast on a test flight and returning to land at Farnborough. Because of the low altitude of the protective 'woolpacks' he flew at only eleven hundred feet. Crossing the coast at Ramsgate, he set course for the Scheldt estuary. They found ideal cloud cover, and between the clouds perfect photographic weather over Flushing and Ymuiden. They flew right across both targets with all five cameras running, then turned back for Farnborough, where the photographic section worked like heroes, well into the early hours.

Kelson drove his master back to Arlington House in time for him to get two hours sleep and an hour for bath and breakfast before that meeting. Peck opened it promptly at ten o'clock, and there was a repeat performance of the day before, a lot of senior officers prepared to put Mister Cotton in his place. After listening for half an hour Cotton opened his briefcase, took out the album of photographs, and asked, 'Is this the kind of thing you want?'

Air Vice-Marshal Peck examined each print, and praised them all. They had been enlarged to twelve-inch squares and each was covered with a transparent Kodatrace on which place names and other details were marked.

'These are first class, Cotton,' he said. 'But we wouldn't expect this sort of quality in wartime.' He handed the album round. One of the officers looking at it asked when the pictures had been taken.

'At three-fifteen yesterday,' Cotton said.

For a few seconds the others registered incredulity. Then, according to Cotton,[6] there was 'indignation as the truth went home. The commotion rose to such a pitch that I wondered what crime I could have committed. "You had no right to do such a thing . . . flouting authority . . . what would happen if everyone behaved like that? . . ." Somebody even said I ought to be arrested. I could stand such nonsense no longer, and decided to get out of the room before I told them what I thought of them . . . I walked slowly to the door and slammed it as I went out.'

Next morning a conciliatory Air Vice-Marshal Peck telephoned Arlington House. He wanted to know when it would be convenient for Sir Cyril Newall, Chief of Air Staff, to call on Cotton. Cotton replied that he could not possibly put Sir Cyril to such trouble. He would call on the CAS at the latter's convenience, and on his way there he would drop in on Peck.

'Would you be prepared to help, Cotton?' Newall asked him over luncheon at the United Services Club.

'Of course I would.'

'Then take charge of the RAF's photographic section.'

'That wouldn't work, sir. The regular officers would resent my intrusion, they already do.' Cotton gave his opinion that that side of the RAF's work was still unsatisfactory because Lord Trenchard had insisted that there should be no specialists in the service, and because the professional aircrews had regarded photographic reconnaissance as unadventurous. He insisted that the best solution was to allow him to form a special unit and to give him the necessary aircraft, and processing facilities. He would like to start at Heston, with his present nucleus of picked men.

'But Heston is a civil airport.'

'Exactly, sir. Nobody would suspect that secret work would be done from there.'

They commissioned Cotton as a squadron leader with the acting rank of Wing-Commander. He was quite used to uniform, and the RAF one did not worry him at all. The new unit was to consist initially of Cotton, five officers, and seventeen other ranks. Cotton requisitioned at Heston the hangars and offices of Airwork, the flying club, and part of the Airport Hotel. Shorty Longbottom, just home from Malta, with Bob Niven and himself made up the 'planning triumvirate'. At the beginning the section depended on the developing and processing department at Farnborough, but Newall had agreed that a photographic section should be built at Heston. Cotton's contact with the CAS was through Peck, and his contact with Peck was through Winterbotham.

Cotton now turned to the question of aircraft, and once more those savage Australian drums beat, and the fray was joined. For Cotton was determined that his section must have Spitfires, stripped, unarmed, polished, and tuned so that nothing aloft could catch them. The experts told him that Spitfires were not suitable and that, in any event, they were unobtainable for such a purpose. In the meantime Cotton had damned well better make do with the two long-nosed Blenheims they delivered to him. Cotton knew they would not do, but he flew the two aircraft to Farnborough, where the Chief Superintendent, Mr A. H. Hall, was most interested in his ideas. Cotton's modifications increased the Blenheims' top speed by eighteen knots, but he still said gloomily that it was far too slow. However this 'Cottonising' of the Blenheims got to the notice of Sir Hugh Dowding, now C.-in-C., Fighter Command. He had to use Blenheims as long-range fighters, and knew them to be desperately slow for that task. Dowding turned up at Heston to satisfy himself that Cotton really could make a Blenheim (or anything else) go much faster. And when the Air Ministry had told him they had no facilities for 'Cottonising' his Blenheim fighters, Cotton promised that if the Air Marshal would requisition another hangar at Heston he would be able to 'Cottonise' Blenheims at the rate of eight per week. He kept his promise. 'We painted the Blenheims the same pale green as my Lockheed,' Cotton says, 'and this became standard camouflage for RAF fighters.'[7]

The grateful Dowding invited Cotton to take tea with him at Bentley Priory, a hospitable gesture fraught with a danger that was amply confirmed. Cotton left the tea table with the promise that two Spitfires would be delivered at Heston at nine next morning. *Then* there was trouble, and not only for Cotton. A mark was chalked up against Dowding (who was to be wrenched from his command immediately after winning the greatest battle in English history). As for Cotton, he was told that he had had no authority to get new aircraft, particularly Spitfires. 'How was I going to service them?' they demanded. 'They needed trained Rolls-Royce staff. Did I intend to pinch those as well? (He did.) His official reply was that his unit was attached (mercifully) to Fighter Command, and that the reallocation of Spitfires was a domestic matter inside the Command. In any event, while the row blazed on he had already converted them, and merely by disarming them, Cottonising them, and polishing all external surfaces into a hard gloss had increased their speed from 360 to just under 400 mph. He next intended fitting a 30-gallon tank under the pilot's seat to increase the range to 1,250 miles at thirty thousand feet. The RAE experts said this would shift the centre of gravity too far aft. Cotton produced his calculations. They compromised on a 29-gallon tank. ('I let them get away with the odd gallon.') He intended fitting cameras weighing 64 lbs immediately astern of the new tank. The same centre-of-gravity argument was produced. Where aircraft were concerned Cotton was an impossible man to argue with. He knew too much.

Calling for a screwdriver, he opened an inspection panel in the rear end of the Spitfire's fuselage and showed the inspectors lead weights amounting to 32 lbs. These had been put there by the manufacturers to counterbalance the extra weight of the new three-bladed steel air-screw. The cameras were fitted and worked well.

At last he was ready to begin taking pictures, though with only two Spitfires, one of which must be used for training a unit destined, he knew, for expansion.

## NOTES TO CHAPTER 12

1. Barker, *Aviator Extraordinary*, p. 103. Soon after the end of the Second World War the author met Cotton, and heard his story in some detail. But nearly all the material in this chapter and 14 is taken from Cotton's own account of his life as told to Ralph Barker.
2. Barker, *op. cit.*, p. 104.
3. Barker, *op. cit.*, p. 118.
4. Barker, *op. cit.*, pp. 32, 33.
5. Barker, *op. cit.*, p. 151.
6. Barker, *op. cit.*, p. 153.
7. Barker, *op. cit.*, p. 158.

# 13
# A Line of Four-Posters

*1939/41*

Bomber Command had learned that, until long-range first-line fighter aircraft could be produced as escorts, daylight raids on Germany were too costly. For the pre-war designers' fancy that bombers with powered gun-turrets would be able to defend themselves in daylight by flying in formation and giving mutual support with cross fire had to be discredited in England, Germany, and America. Eventually all three knew that, in the face of strong fighter opposition, such tactics meant death.

It is an interesting point, proving that designers, as TRE tried to do, should work with the airmen or soldiers or sailors for whom they design, and that they should work under war conditions. Between the wars the bomber designers saw their aircraft as swift, defendable 'flying fortresses' (to steal an American proprietary name). These aircraft were splendid on paper. They captured the imaginations of politicians and flying men. But the designers had not foreseen how rapid was to be the improvement in fighter aircraft or, in the inverse, how dreadfully the bomber was to be handicapped by its size and its weight. The fighter pilot, secure in his speed and balance, aimed *his whole machine* at the target, and hose-piped into it with observable tracer. The gunner in a powered turret was infinitely at a disadvantage. He was more vulnerable, he was probably much more uncomfortable —sometimes almost frozen to death—and his mentality was that of the defender rather than the attacker. In addition the turret gunner had (as any civilian shooting sportsman will readily understand) immensely difficult problems of skill and prognostication. Supposing, for the sake of simplicity, that his aircraft and that of the enemy were both flying at three hundred miles per hour, he might, though this was unlikely, have an easy broadside target, or he might, and this was more probable, have a target coming at him at an angle, and at a speed approaching six hundred miles per hour. . . . No, the solution with the bombers was to achieve air superiority and to give them fighter defence. That was what the Germans were able to do at the beginning. That was what the Allies did in the latter stages.

For the English the lesson came soon after midday on December 18, 1939. Twenty-four Wellingtons were on patrol near Wilhelmshaven with orders to bomb any German naval unit found at sea. (Bombing of the German mainland was at that time prohibited by Whitehall.) The *Freya* on Wangerooge had picked up the Wellingtons at a range of more than seventy miles and put out the alert. German fighter pilots were already sitting down to lunch, but as the Wellingtons neared the *Freya*, fifty fighters, Messerschmitt 109s and 110s, climbed above them into the cloudless sky. Maintaining their diamond formation, the Wellingtons turned for England and the cannon-firing Messerschmitts swooped on them. The British lost fifty-eight per cent, since only ten Wellingtons got home. Strangely enough, the Luftwaffe did not learn from that afternoon's work; they did not learn until August 15 in the Battle of Britain, when Stumpf sent his unescorted bombers to England across the North Sea.

So, with such resources as it could muster Bomber Command attacked by night, when Churchill, in May, 1940, had lifted the ban on bombing Germany. Within a month more than two thousand bombing sorties were flown with minimal losses. By the standards that were later to prevail, the raids were small and inaccurate because the British as yet had no radio bombing aids. Nevertheless the attacks had a disproportionate effect in Germany. Firstly, they constituted an unpleasantness that was plainly going to get worse before it got better. Secondly, Göring himself had declared that not a single bomb would fall on the Ruhr! *Now*, Göring began to demand more D/T production; and the same kind of helplessness under night bombing that the British had experienced was endured by the Germans. Something had to be done, and Colonel Josef Kammhuber, aged forty-three, was called in to do it. Göring promoted him to Major-General, told him to take twenty-four hours or so to study the matter, and then get on with building an effective defence. He would have priority in D/T apparatus, searchlights, guns, and fighter aircraft.

Kammhuber considered the situation as he found it. The types of machines used as night fighters were the Me 110, the new fighter version of the Ju 88, and some Me 109s. Both twin-engined fighters were useful gun platforms for dealing with slow bombers, but the fast mono-motor 109 was making the majority of the kills in the 'Helle Nachtjagd'—floodlit night hunt. The fighters waited until there was a warning from the *Freya* outpost line. They then orbited their airfields' radio beacons until the German searchlights and flak, or the fires started by their bomber foes, illuminated or silhouetted the bombers: then they closed and attacked. The stupidity there, Kammhuber saw, was that the enemy had probably already dropped their loads; and the German night fighters were operating in their own flak areas.

His answer was to lay a defence line across British entry to Germany, running from the northern tip of Denmark to the Elbe estuary then west by south along North Germany and Holland, and then down through Belgium and France to the Swiss–Italian frontier. The line was to consist of a series of equally-spaced defence stations or radio boxes, each box with sides some thirty kilometres in length. The code-name for such a radio box was *Himmelbett*, four-poster bed. By putting four-posters on the Friesian and other islands he planned from the start to strengthen the line at crucial points and deepen it strategically. He increased the number of early-warning stations along the coasts and he also, in collaboration with General Martini, obtained a build-up in the German Signals' monitoring service. This proved to be invaluable when faced with so intense, and at first indiscriminate, a user of radio as RAF Bomber Command.

Because the defence operated in a series of identical boxes, it is fairly simple to explain. Each box had two *Würzburg* radars and one *Freya*. They were sited in a triangle with sides of less than one kilometre. In the centre was station headquarters and the control room with its two-tier Seeburg Table. What happened was this:

Early warning came through that RAF bombers were approaching.

One night fighter (in each box) took off and orbited the *Himmelbett*'s radio-beacon, in contact by radio-telephone with the *Himmelbett*'s controller. The German fighter was followed by the station's 'blue' *Würzburg*.

The station's *Freya* picked up some of the raiders which, in the early stages of British night bombing, as in the German bombing of Britain, were dispersed and travelling individually to their target(s).

If a British bomber came towards or near the *Himmelbett* the 'red' *Würzburg* helped by the *Freya* fixed on it while the 'blue' *Würzburg* continued to track the night fighter. Each of the two *Würzburgs* was connected by direct line to an operator inside the lower (ground level) stage of the station's Seeburg Table. The Seeburg was a two-storeyed structure made of wood that represented on a small scale the whole cube of that particular *Himmelbett.* One operator at the table operated the red (bomber) *Würzburg* simulator while the other operated the blue (fighter) one. The *Würzburg* simulators were at floor level and their red and blue pencils of light came through the table and shone up, a small blue circle and a small red one on the ground-glass 'ceiling' above. At the upper level the fighter controller stood, looking down on the ground-glass gridded plan of the sky, the 'roof', so to speak, of the *Himmelbett.* He was in constant communication with his fighter pilot on the radio telephone and while watching the blue blob and the red jolting across the grid, talked the fighter into contact. If the fighter pilot failed to find his prey he returned to orbit round the radio beacon. It was a simple and ingenious scheme.[1]

As his *Himmelbett* line grew and with practice the pilots and controllers made speedy improvements, Kammhuber altered the siting of the German searchlight belt, to give a greater chance of catching a bomber that had slipped through. The belt was made out of bounds to all German aircraft save night fighters. In the autumn of 1941 the searchlights were removed by the order of the OKL and taken from Kammhuber's command. He tried to prevent it, but when the searchlights had gone he found that the proportion of kills rose, because the pilots ceased to rely on anything other than Radar and their controllers' vectors.

Kammhuber's main concerns were firstly to improve on the *Würzburg* and secondly to get a workable Air Interception Radar in his night fighters. The efficiency of each station in the line depended on its *Würzburgs*, because the *Freya* did not give the altitude of the incoming bomber. The *Würzburg* did, but its range of twenty miles was too short and it was liable to suffer from ground reflections if the bomber flew lower than six thousand feet. Kammhuber lost no time in throwing these defects at the manufacturer. Unlike Britain, Germany still had plenty of slack to take up in her manufacturing industry, and Telefunken were prompt with an improvement. Before the winter of 1941 they were experimenting with a new Giant *Würzburg*, and very shortly this began to replace the standard machines inside the *Himmelbett* boxes. The Giant was not readily transportable. Its bowl-shaped skeletonic reflector was twenty-five feet in diameter as opposed to its small parent's ten feet. The Giant threw a narrower beam, and it had double the range—forty miles.

As to airborne Radar, the British had a long start, and also, though the Germans did not yet know it, a tremendous superiority with their new valve, but Telefunken did wonders. By the end of January, 1942, just before the Bruneval Raid, four Junkers 88 night fighters were carrying the new *Lichtenstein* set. The *Lichtenstein* was heavy, and operated through a gruesome jut of aerials on the Junker's nose, but it had a maximum range of two miles, similar to that of the early TRE models, and a very good minimum range of two hundred metres. There was strong pilot resistance to the *Lichtenstein*; this was partly because there was great confidence in the *Himmelbett* system of controlled interception, and partly because the proliferation of aerials all but ruined the fighter's handling and performance.

So the situation was that the British were building up their bombing campaign, and their aircraft industry was turning out, particularly in the Lancaster, better long-range bombers than the Germans had or would have, but Bomber Command had still to apply science if its night bombing was to be anything but inaccurate.[2] Meanwhile Germany, although engaged on several fronts and particularly in Russia, where the initial thrust had gone in very deep, was reacting with speed and brains to the bombing threat. German fighter production, spurred by the obvious need for night fighters, had at once risen dramatically, and there was scope for a continual expansion (which materialised until the end of the war). The British were already losing night bombers in the order of about four out of every hundred. Some of these losses were attributable to the German flak with its (*Würzburg*-directed) blue master-searchlights, but at least two-thirds of the victims were, it was learned in de-briefing, shot down by night fighters.

* * *

If the Germans were not using Radar, how were their night fighters managing to do so well? Dr R. V. Jones had not forgotten the 'radio detection stations along the North Sea Coast' mentioned in the Oslo Report. He, like the German-trained Professor Lindemann, kept an open mind, but it was difficult to find anybody in England willing to consider the possibility that Germany too might have Radar. Now British aircraft were falling over Germany, and it was the Germans, rather than Jones and his Intelligence colleagues, who had the advantage of crew interrogation. British Radar had involved the erection of huge masts and towers; if the Germans did possess Radar they must have another variety that did not require such formidable structures. Jones noted that in May 1940 a prisoner had mentioned a radio gun-laying and ranging device used by his Navy and had also spoken of the Luftwaffe's radio warning system. Then, in July, 1940, an Intelligence source in Northern France sent over a section of a German report in which a '*Freya* warning' was mentioned, and the fact that it had been able to put German fighters on to intruding British aircraft. Jones had at once asked for any further information on '*Freya*' to be given priority. In due course a report came of a *Freya* station that had been set up at Lannion, a small port in the north-west corner of the Côtes-du-Nord Department. Jones could think of no direct significance in Lannion itself, and there were no reports of secret works there; but an early-warning station there would make sense, Lannion being on an air route to Brest and the Biscay ports of France, which the enemy would obviously turn into his most important submarine bases. It *must* be significant that the Germans had established a *Freya* there only three weeks after entering France. Also, according to the agent, it was under a twenty-four-hour armed guard and had its own flak protection. What then was a *Freya*?

Turning back into mythology, Jones noted that the goddess Freya had betrayed her husband in order to possess Brisingamen, a necklace. The magic necklace was guarded for Freya by Heimdal, servant of the Gods, who could see for a hundred miles in every direction, in daylight and in the dark. Freya? Would they have been so obvious as thus to name a Radar device? Then, it must be comparatively small. The air photographs of Lannion showed nothing significant. He wrote a memo to the Prime Minister.

'Heimdal himself would have seemed the best choice for a code name for RDF,' Churchill read. 'It is difficult to escape the conclusion that the *Freya-Gerät* is a form of portable RDF. Freya may possibly be associated with Wotan—she was at one time his mistress—although it would have been expected that the Führer would have in this case chosen Frigga, Wotan's lawful wife.'

Churchill at once required General Ismay to ascertain if any RDF sets had fallen intact into German hands at the time of Dunkirk. The gist of Ismay's findings was that one Radar set had been left behind by the RAF but that it had probably been destroyed. The Genera pointed out that the Germans might have obtained very complete information about RDF from the French, to whom all things, even RDF, had been vouchsafed before the débâcle.

A further Intelligence report from France soon reached Dr Jones at the Air Ministry. Another *Freya* station was definitely located on Cap de la Hague, north-west of Cherbourg. On July 23 (1940) this *Freya* had guided German dive bombers to the destroyer HMS *Delight*, which they sank, a good sixty miles from Cherbourg. Visibility was thick, and *Delight* had done nothing to reveal her position. It looked as though *Freya* was an efficient long-distance instrument. From Paris, too, Jones had been sent a copy of Daily Orders at the headquarters of the Third Luftflotte. *Freya* was mentioned. It was a part of German air defences, and must exist in large numbers.

German Radar was first heard and recognised in England because Jones was on another of his radio-navigational trails. He had always wondered if the Germans, like his own people, were working on centimetric techniques, and he therefore paid close attention to evidence from the Continent that a German apparatus existed called *Knickebein Dezi*. The standard *Knickebein* worked on a wavelength of some ten metres. Was it not reasonable to suppose that *Knickebein Dezi* might be a similar thing but working on a wavelength of a few decimetres? Further, he heard of a German technician making an alteration of thirteen centimetres in one of the beams.

'I was able to get Prime Ministerial pressure applied to a search at these wavelengths,' he says.[3] 'Although I myself regarded this particular deduction as a slender one, I was convinced that a search ought to be made on general principle. As it turned out the deduction was false; *Knickebein Dezi* certainly proved to be a decimetric installation associated with the *Knickebein* system. But it was the communications link which tied the system into the German signals network. . . . Moreover the mention of thirteen centimetres had referred simply to the moving of a monitor aerial used as a sitting target for directing the beam. The radio search which was instituted on these false premises, however, proved to be fully justified. . . . It revealed to us that the Germans had a Radar system surveying the Straits of Dover. . . .'

Derek Garrard, who had been working on Radar at TRE, was to be attached to Dr Jones's staff at the Air Ministry. As part of the search mentioned above he was encouraged to take a few days off between

Swanage and London, and to do some beam hunting on the short wavelength. He packed a lot of TRE equipment into the back of his car, and drove east along the chalk downs of Southern England, doing spells of radio search. From near Dover he heard strange signals on 375 mc/s. Further, he discovered that the transmissions were connected with the shelling of British ships passing through the Straits. They were being attacked by the German coastal batteries whose presence annoyed Churchill so much, and the Radar emissions Garrard had discovered were those of the shore-based German *Seetakt*, the naval gunlaying unit. Garrard's discovery was soon being digested at Swanage and in London, and was investigated in every way that could be imagined. It made people uneasy. They would have been yet more uncomfortable had they realised that *Seetakt* was operational during the Spanish Civil War, and that a report on its aerial array had been in Admiralty files for eighteen months. Naval Intelligence had noted during the *Graf Spee*'s tour of duty in Spanish waters a large shrouded aerial above the bridge. When, after a running battle with three British cruisers, the German captain saw fit to scuttle his pocket battleship in the shallow mouth of the River Plate, a British Radar specialist, L. Bainbridge Bell, joined the sightseers who went out in small boats from Montevideo to examine the stranded and damaged monster. Bell had climbed up to the aerial, and reported to London that it almost certainly belonged to a Radar unit used for ranging the battleship's guns.[4]

It began to seem to scientific intelligence that the Germans not only had Radar, but sophisticated Radar. If so, the British were working in a dangerous vacuum. But now a new aid was coming to hand in the search for the elusive Dezimeter Telegraphie; the aid was an improved kind of aerial photography devised by an Australian.

## NOTES TO CHAPTER 13

1. For an excellent description of Kammhuber and his system of defence see Price, *Instruments of Darkness*, pp. 63–70.
2. Harris, *Bomber Offensive*, pp. 80, 81.
3. Jones, *Minerva*, Vol. X, No. 3, p. 443.
4. Price, *Instruments of Darkness*, p. 17.

# 14
# The High-Altitude Handshake

*1940*

Sidney Cotton of course lost no time in getting to work from French bases. If his relations with the Air Staff continued to be strained, he was soon on excellent terms with Air Marshal Sir Arthur Barratt, who commanded the RAF in France. There was an immensity of work to be done on the German frontier whenever the weather was suitable, and as the Belgians (afraid of annoying the Germans) refused to allow Anglo-French aerial reconnaissance over their territory, and the Belgian maps were considered inadequate by Lord Gort and the staff of his British Expeditionary Force, Cotton was asked, unofficially, to get a complete aerial cover of the country. One thing about 'that fellow Cotton', the more he was asked to do, especially if it seemed a little contrary to the law, the more eager he was to tackle it.

Film poured in from Shorty Longbottom, who did many of the initial flights from foreign fields, and Cotton had bother in getting it interpreted and processed. 'By the end of the war,' he says,[1] 'the Allied Central Interpretation Unit employed 550 officers and 2,000 other ranks, providing 80 per cent of our intelligence on the enemy. At the outbreak of war our interpretation capacity consisted of two RAF officers at the Air Ministry, two Army officers, and a small nucleus at the Admiralty.' Nor would the Air Ministry interpret his film readily, since although its quality was good it was taken from thirty thousand feet instead of the ten thousand usual with the RAF and the French Air Force. Cotton insisted that the task of maintaining a watch on Germany *must* be done from thirty thousand, and that reconnaissance from ten thousand was suicidal. 'We must learn to interpret from that height. And we must get better cameras.'

He found an unofficial way to get interpretation, and at first he financed it out of his own pocket. 'Lemnos' Hemming (he had lost an eye in an air accident) had worked with Cotton in Newfoundland,

and was now running an aerial survey business at Wembley called the Aircraft Operating Company. Hemming knew that his business would have a valuable part to play in the war, and he had done his best to get it taken over by the Air Ministry, which refused to consider such a step. Cotton took this up with Air Vice-Marshal Peck, but even Cotton failed to convince Peck of the value of Hemming and his trained interpreters who worked stereoscopically, and used a large and extremely sophisticated Swiss calibrating and measuring machine known as the Wild (pron. Vildt).

Maddened by his growing backlog of uninterpreted film, Cotton privately took over Hemming's firm, swearing every member of it to secrecy. This raised the standard of interpretation and showed that his thirty-thousand-feet cover was clear to trained technicians. Hemming also produced a camera with a twenty-inch focal length, which gave a scale of 1/18,000 at thirty thousand feet, as against 1/72,000 for the five-inch camera. Cotton's unit successfully used both cameras until the Air Ministry produced the F.52 which gave a scale of 1/10,000 from thirty thousand feet and became the standard air camera of the war.

Although during his clandestine beginning Sidney Cotton had not worked readily with the Deuxième Bureau, he now found the French helpful. General Vuillemin, French Chief of Air Staff, in return for certain photographic favours, gave him access to his airfields. Cotton found these more secure and more comfortable than RAF fields in France, where station commanders might resent the duck-egg green Spitfire, Cotton's Lockheed, and his perfect manservant driving a large private car with an illuminated Union Jack on the back. Vuillemin had a special hangar built for Cotton at his own field, Coulommiers. It could house the Lockheed and two Spitfires and was roofed in straw thatch to resemble a haystack. It escaped, as few hangars did, the German attacks on French airfields in May 1940.

RAF and French aircraft had been trying for weeks and with severe losses, to photograph the Ruhr. Bob Niven stood by for days waiting for a clearance over that smoke-palled area until on December 29, in less than half an hour of Spitfire flight, he covered the whole southern half. During the next few clear days he photographed Cologne, Dusseldorf, and much of the Siegfried Line. Even this, and loud and forceful were Cotton's plaints, did not induce the Air Staff to keep their promises to him. He should by now have had eight Spitfires, a trained RAF staff officer to take some administrative weight off his shoulders, two photographic trailers to work with his section in France; and, most of all, he wanted Hemming's AOC to be officially recognised and incorporated with his Photographic Development Unit. But he now had help from an unexpected quarter. He was introduced to Colonel Lespair, who commanded the French School of Photographic Interpretation at Meaux. 'I was shown beautiful dossiers of photographs. It was plain from the detailed annotations and analyses that the French were a long way ahead of the RAF in interpretation and that even Lemnos's men could learn a lot from them.'[2] He arranged for one of 'Lemnos's men', Douglas Kendall, to be allowed to take the French course (and Kendall eventually became one of the leading photographic interpreters of the war). Meanwhile Cotton also used the French school to process some of his unit's film. Lespair was at first astonished, then fascinated, by Cotton's theories; and when he heard of Cotton's difficulties with the Air Staff, Lespair produced a chart which (summarised in words) showed:

(i) The RAF had photographed twenty-five hundred square miles of German territory in three months for the loss of forty aircraft.

(ii) The French Air Force had photographed six thousand square miles for the loss of sixty aircraft.

(iii) Cotton's unit had photographed five thousand square miles in three flights without loss. (And these figures did not include twelve thousand square miles of Belgian territory which the unit had secretly photographed at the request of Lord Gort. That also entailed no loss.)

Cotton flew Lespair's figures to Peck in London. He guessed as he watched Peck studying the paper that the figures were 'about right'. He was taken straight in to see Air Marshal Peirse, and the following day he was told that the Air Council had approved the expansion of his unit. His reply was tart; it had been approved much earlier by the Chief of Air Staff. Nine days later the first of his new Spitfires touched down at Heston. He made another vain attempt to get Lemnos Hemming's company recognised and to get the administrative officer that, with his business man's longing for clarity and order, he craved. However, Air Marshal Barratt soon produced the latter, Geoffrey Tuttle, one of the best officers on his staff in France. This was a momentous appointment.

Cotton's next real tussle with the Establishment came when Naval Intelligence got hold of a report that the *Tirpitz* had put to sea, and the RAF had failed to get photographic or visual proof, one way or the other. Commander Charles Drake of the NID telephoned Cotton about the *Tirpitz* on February 6. The weather was thick then, and for three days. But at eleven o'clock on February 10 Shorty Longbottom took off for Wilhemshaven. Cotton telephoned Lemnos Hemming to keep some of his staff on duty all night.

'It was a cold, clear day. The Spitfire, stripped of all armament, streamlined to our pattern, and polished till it shone, climbed rapidly away from Heston. At three-twenty that afternoon Shorty was safely back, having photographed Wilhelmshaven and Emden. . . . I took the film straight to Hemming at Wembley but the special processing for ship recognition took some hours, and it was two on Sunday morning before the film was ready for interpretation.'[3]

Michael Spender, brother of Stephen Spender the poet and of Humphrey Spender the painter, was the interpreter of ships. He was a Clarendon physicist, like other distinguished persons in this book. He had joined Hemming's firm in 1939, and when it came to work under Sidney Cotton he was one of the first to grasp that air intelligence could inform in time as well as in space. In other words if the Spitfires could photograph Kiel regularly, and one cover was compared with its predecessors and its followers, the interpreter should get a good idea of what the enemy was *doing* there, and what might *happen* there —not just an idea of what *was* there. Photographic intelligence must be read in depth of scene, context, war background. Then the reader may penetrate the enemy's thoughts and know his plans.

That Sunday morning in the early hours Spender was able to state that the *Tirpitz* was still in graving dock. It seemed essential to Cotton that a Royal Navy specialist should also, and at once, examine the cover. He telephoned the Air Ministry and was told to take the processed film there. He must not contact the Admiralty. Cotton drove straight to the Ministry and delivered a complete set.

By Monday evening the Admiralty were getting desperate for information. They had not yet been shown the pictures, and had heard from the Air Ministry that there were sixty submarines at Emden. Michael Spender had (correctly) recognised these as a group of barges, not submarines. Cotton decided that 'the play had gone on long enough', and sent to Naval Intelligence a set of prints with Spender's interpretations on the traces. The Admiralty realised that it could have had the correct interpretation thirty-six hours earlier. The result was that Churchill listened with mounting wrath to the complaints of Sir Dudley Pound, and Cotton was ordered to attend the War Room meeting that night. Mr Churchill intended to be there to sponsor him. But he was not; he had been summoned unexpectedly to Buckingham Palace, and Cotton faced the situation of a bushel of grain being carried towards two millstones. To make things worse, Pound settled himself in his high-backed chair and motioned Cotton brusquely to take the similarly high-backed chair on his right hand, the place of honour normally occupied by Air Marshal Peirse. Peirse had at once enquired on entering the room, 'What are *you* doing here, Cotton?' Then Admiral Pound had intervened. 'Sorry about this, Peirse, but we sent for Cotton in rather a hurry, and there wasn't time to get permission from Air Staff. We particularly want him here tonight.'

Shorty's pictures were passed round with the annotations by Michael Spender including the Wild measurements of the German ships (and barges). Peirse asked how these had been arrived at. Cotton answered, 'With a special photogrammetric instrument, sir.'

'Why weren't Air Staff told about it? We'd have requisitioned it at once.'

Cotton says that he had in his pocket a copy of the Air Ministry's latest refusal to do just that, but he kept quiet with an effort, and 'accepted the rebuke'.

Sparks then flew between Pound and Peirse. Cotton had never been averse to sparks. He had always got on well with the Navy because they appeared to accept him as a person in his own right, and were not disconcerted by his lowly official rank; then, while his work was invaluable to them, he did not have to pester them on administrative matters. According to Cotton,[4] the next part of the meeting went like this:

POUND We have pressed for information on the whereabouts of those ships and have been put off for one reason or another. We are not prepared to accept that situation any longer.

PEIRSE I thought we had made it clear that this is a most difficult task, and that some of the best brains in the Air Force are working on it.

POUND Perhaps you'd get better results if you tried some of the lesser brains. . . . (turning to Cotton) Cotton, you've heard what we've been discussing. Can you get us this information?

COTTON Yes sir. Quite easily.

PEIRSE (Jumping from his chair) I do not accept that, just because Cotton says it. There are scores of difficulties he probably hasn't even thought of.

COTTON Surely the proof of the pudding's in the eating, Air Marshal. May we not try?

POUND Sounds reasonable to me. What about it, Peirse?

PEIRSE How do you propose to carry out this operation? I shall want to know a lot more about it before I'm convinced it's feasible.

Some technical argument followed between Cotton and Peirse and then the latter arranged that they would 'work out the details' the following morning. 'And bring me the information on the equipment you mentioned earlier,' Peirse enjoined.

'Pound leaned across at the Air Marshal. "Are you going to requisition that equipment *now*, Peirse? Because if not we propose to do it ourselves." And the Air Marshal answered, "Yes, I should have been told about it before." '[5]

* * *

A week after Cotton's little victory in the War Room (which was not in the end a victory except for Hemming and the Wild machine) he heard from the Admiralty that cover was urgently needed on the German ports. The Air Ministry told him they were making regular flights over Heligoland, and obtaining the necessary information. Checking back with the Navy, Cotton found they had only been given visual information obtained by moonlight. He also learned that in carrying out those sorties three Blenheims had been lost out of five. 'I regarded this slaughter as unnecessary and little short of criminal,' Cotton says. And he said so openly, that being his nature. He could not help 'exploding', as he put it, at the 'shocking waste' of sending

Blenheims on tasks for which they were not fitted, and the 'criminal aspect' of losing splendid aircrew on such unaccomplishable tasks. At the Admiralty's request he went on photographing Heligoland and all other German targets given him by the NID.

It was now his turn to attack the Air Ministry. Earlier he had recommended Bob Niven and Shorty for the Distinguished Flying Cross. The Ministry's press release on the announcement of the awards said that the pilots had been engaged on developing a new kind of photographic intelligence. Cotton naturally felt that this was 'a disturbing and annoying lapse of security'. He had come to believe 'that jealousy of my unit's reputation was mounting. I was aware that there were men in the Air Ministry intriguing against me . . . but apart from repeated requests for speeding up the expansion of my unit I took no action except to appeal to senior officers not to listen to gossip, but to back us up fully.'[6]

On March 2 Niven took off in a Spitfire only delivered a few days previously from Supermarine. It was fitted with Cotton's new wing tanks and had a range of two thousand miles. From thirty thousand feet Niven took a complete set of pictures of the whole Ruhr. At the end of his traverses three German fighters rose to intercept near the Luxembourg frontier. Niven opened up the Merlin to maximum permissible revs. It seemed to him that he just slid away from the three pursuers, who might have been flying in glue. After half a minute he cut back his revs to save petrol. Niven's pictures were so good that Cotton had them made into a single mosaic. When he opened this out in front of Air Chief Marshal Sir Edgar Ludlow-Hewitt, C.-in-C. Bomber Command, he in his turn put forward a request that Cotton's unit should belong to Bomber Command. This Cotton opposed, on the grounds that 'it was unwise for a major customer to run the show'.

On the day following Niven's triumph Cotton experienced his Black Sunday. 'First, one of my new pilots crashed our training Spitfire, which seemed tragic enough at the time. Then real tragedy overtook us with the news that Dennis Slocum's Hudson had been shot down over Kent by Spitfires from Biggin Hill. . . . Slocum was killed, and so was the wireless operator lent to me by Ludlow-Hewitt.' One of Cotton's more delightful characteristics was his sense of identification with his pilots. 'Slogger' Slocum, a former airline pilot, first flew in Cotton's Circus as a free-lance on short leaves from his own squadron, which was stationed in Scotland. When he managed to get transferred, he came up to even Cotton's standard. On his first long-range flight, over naval targets including Cuxhaven and Brunsbüttel, his cameras caught German submarines surfaced. The pictures were so remarkable that the Admiralty decided to show them to the King. 'Cotton could always pick 'em,' his unit used to say. They also picked Cotton, as was shown by Slocum and others like him.

That March, Cotton was given sensible advice by three high-ranking airmen, Ludlow-Hewitt, Barratt, and Air Commodore Douglas Colyer, then Air Attaché in Paris. He had earlier had the same sort of advice from Tedder and many another. They all told him to stop fretting himself into a frenzy at the slowness of the Air Ministry, to stop tilting at windmills, to appear to obey all orders, and at the same time to push his unit ahead quietly. He could not follow such advice. His frustrations boiled over, and as he had said to the Chief of Air Staff, he was an Australian, and talked not of spades but of bloody shovels. He enjoyed taking the Admiralty line, for example, on the Radar defence system. He carried no IFF (Identification Friend from Foe) in his Lockheed 'simply', he says,[7] 'because no one had ever suggested it'. While admitting that his 'White Flight Taking Off' warning had always been given to Fighter Command and that his Lockheed was a civilian aircraft and easy to recognise, he felt that if the early-warning system worked at all he should have been buzzed quite often by British fighters. His attitude to British Radar was more deprecatory than Churchill's 'M'yes', and he was unwise enough to use the Lockheed in Admiralty probes to test the system. To the Air Staff his conduct there seemed disloyal. The truth probably was that the Chain Home screen picked up the Lockheed, but the air defences were not set in motion for one isolated aircraft that in any event would be recognised by the Observer Corps for what it was. Later in the war the Tempsford Squadrons which did the parachuting and clandestine landings in occupied Europe were to prove that single bombers could fly through (or under) the very effective German Radar screen, do their work, and return unchallenged.

Interpreting for the unit in France on May 7, 1940, Douglas Kendall (his French training completed) saw from the record of a Spitfire sortie that there were German tank units hidden in the Ardennes forests. Cotton took the still-wet negative to Air Marshal Barratt, who asked for an immediate low-level sortie. This showed that there were some four hundred German tanks *visible* in quite a small area. Barratt was alive to the significance of the discovery. The Allies were expecting the main German attack to come across the Belgian plain, and it was generally believed (though some experts ridiculed the belief) that the German armour could not pass through the Ardennes.

Unable to get any reaction that he considered strong enough from London, Barratt asked Cotton to take his photographs to the C.-in-C. Bomber Command. If the British 'heavies' (there were now some two hundred of them) bombed that section of forest with incendiaries and high explosives, the enemy's petrol and oil munitions dumps as well as his armour would certainly be at risk. Cotton flew himself to Heston and drove without delay to Bomber Command Headquarters at High Wycombe. His friend Ludlow-Hewitt had been replaced by Air Marshal Sir Charles Portal.

Although Cotton was an unofficial emissary, Portal received him. 'The interview proved an unfortunate one,' Cotton says.[8] 'He clearly regarded me as a nuisance. . . . I showed him the photographs, but I could see that my task was hopeless.' Cotton knew, of course, that Barratt had no power to call in the heavies, though when real war began he would be able to ask Bomber Command for assistance. Yet what he asked for now was a bombing attack on a German target before the Germans had actually started shooting. Even had Cotton convinced Portal and had Portal convinced the War Cabinet there would hardly have been time. The Panzer divisions, seven of them, began to roar out of their forest base early the following morning, May 10, and their thrust pulverised Europe.

With the outbreak of the shooting war Cotton was told that the Air Ministry had decided to expand his Photographic Development Unit at an unprecedented rate. But on May 16, when the Dutch had surrendered, and Churchill had flown to Paris to see Paul Reynaud, Cotton, who was in London, learned that the unit had been ordered to leave France. 'Naturally, though, they were waiting for my orders.' He signalled them to stop where they were, pending his arrival, and piloted the Lockheed over to Meaux. He was soon with Air Marshal Barratt whom he found 'like everyone else, looking very tired'. Cotton persuaded Barratt that now, more than ever, he needed the services of the only reconnaissance unit that could and would survive the German onslaught. As for the evacuation of his people and their gear demanded by London, Cotton assured the Air Marshal that with their Hudsons even his ground staff were completely mobile, and he would guarantee their safety, and that none of them should fall into enemy hands. The following day there were more orders from London to leave France. Cotton sent back all the men he could spare, in one of the Hudsons. Meanwhile he and the unit continued to operate under Barratt for a further three weeks. On June 14, when his Spitfires had at last been withdrawn from French bases and the rest of the section were on the move south to Poitiers, Cotton flew to London. Geoffrey Tuttle warned him then that the Air Ministry had belatedly awoken to the merits of his methods, and that they were planning to take over the unit and sack him. Cotton would not believe it. He asked his friends in Air Intelligence if such a thing could be, and they denied it. He was steaming from his activities in France, and was in a hurry to take off for Poitiers, where he arrived that same afternoon. The unit had increased its transport by salvaging an airworthy Fairey Battle, found abandoned. Before he and Bob Niven left for England in the Lockheed, all his people having been safely despatched home, they picked up two stranded waifs, an English girl secretary and her collie. Cotton noted, 'the dog seemed to enjoy the flight'.

They flew into thick fog over the English coast. He accordingly turned back for Jersey and took rooms for himself and his passengers (including the dog) in a hotel, which was machine-gunned by the Luftwaffe early next morning. Expecting German aircraft to be active over the Channel, Cotton filled his tanks and flew out into the Western Approaches before turning up north and coming in over Bristol, and so on to Heston. As he stepped from the Lockheed he was handed an official letter. He was puzzled by the address, since his initials were incorrect and he had been credited with a decoration he had never won. He read:

Air Ministry, Dept. OA
London SW1
16 June 1940

SECRET[9]
S.58864/S.6.

Sir,

1 I am commanded by the Air Council to inform you that they have recently had under review the question of the future status and organisation of the Photographic Development Unit and that, after

careful consideration, they have reached the conclusion that this Unit, which you have done so much to foster, should now be regarded as having passed beyond the stage of experiment and should take its place as part of the ordinary organisation of the Royal Air Force.

2 It has accordingly been decided that it should be constituted as a unit of the Royal Air Force under the orders of the Commander-in-Chief, Coastal Command, and should be commanded by a regular serving officer. Wing-Commander G. W. Tuttle, D.F.C., has been appointed.

3 I am to add that the Council wish to record how much they are indebted to you for the work you have done and for the great gifts of imagination and inventive thought which you have brought to bear on the development of the technique of photography in the Royal Air Force.

I am, Sir,<br>
Your obedient servant,<br>
Arthur Street

Wing-Commander H. L. Cotton, AFC<br>
Royal Air Force Station,<br>
Heston, Middlesex.

Sir Arthur Street was Permanent Under-Secretary to the Air Ministry.

At last Cotton's enemies, while his major ally Air Marshal Barratt was still busy with the French defeat, had thrown him out. The Admiralty, which had been responsible for some of his unpopularity with the Air Council, was too lazy, or too discreet to stand up for him. And when, later, Their Lordships offered him employment they were told by the Air Ministry that they might not employ him, that if they did so it would be considered a hostile act; and they acquiesced. Cotton's efforts to get back into his own unit were politely stifled. Every effort he made to do something to help win the war was smothered by the remorseless and sound-proof blanket that the Establishment can drop over any unwanted or resented person, particularly in time of war. The services of this one energetic, patriotic, and brilliant individual were made tabu by a Ministry that knew his brilliance had made it look foolish. Cotton was a man of many interests, and he did not lie down and die. His consolation was that his unit and the nucleus of officers and men (including Geoffrey Tuttle) whom he had personally chosen, went from success to success. The views he had postulated on high speed photographic reconnaissance and scientific stereoscopic interpretation in depth, views which had been laughed at, ridiculed, resisted, were now the official views. And the unit, now that he who had conceived it and who had raised it up had been sacked, was one of the brightest gems in the Establishment's regalia. . . . England can be merciless to those who serve her, icily polite and unemotional.

## NOTES TO CHAPTER 14

1. Barker, *Aviator Extraordinary*, p. 165.
2. Barker, *op. cit.*, p. 169.
3. Barker, *op. cit.*, p. 171.
4. Barker, *op. cit.*, p. 175.
5. Barker, *op. cit.*, p. 177.
6. Barker, *op. cit.*, p. 180.
7. Barker, *op. cit.*, p. 184.
8. Barker, *op. cit.*, p. 189.
9. This secret classification no longer applies.

# 15
# Stereoscopes and Dicers

*1941*

'Cotton, the late Sidney of that ilk? A born pilot, marvellous pair of hands, but an adventurer, an out and out adventurer. Yes, my dear boy, I know Drake was too. But Cotton was impossible. His own worst enemy. Always narking at authority. I had a good deal to do with him, for my sins, in 1940. His own fellows would do anything for him. But one could never get hold of him, because he was always flying here or flying there. When you command a unit you've got to be chairborne *some* of the time. I always felt he was enjoying the war at the taxpayers' expense.' The speaker was a retired Air Marshal whom I met in his club in Piccadilly twenty-five years after the war, when Cotton was dead (he died in 1969) but not forgotten.

'But he paid for things himself,' I remonstrated. 'The American self-sealing compound for the wing tanks of his Spitfires, for example.'

'Might well have done. Liked splashing money about.'

'Why not? It was his own money. So far as I know he only had a squadron-leader's pay of *your* money. In any event, but for Cotton you'd never have had the PRU, and that *would* have been a loss to the Royal Air Force and to the country.'

'When we needed Spitfire fighters he wangled them for photography. Anyway, it was after Cotton got the push that the PRU came through, under Tuttle, a brilliant officer.'

As Cotton was delighted to declare, Geoffrey Tuttle made a remarkable success of the Photographic Reconnaissance Unit; and he rose to be an Air Marshal and to receive a knighthood. Yet when he met Cotton at any time after the latter's discharge it was the senior regular officer who called the junior reservist 'sir'. Tuttle amply showed that in choosing him Cotton and Barratt had been wise. He cemented and increased the virility of Cotton's unit, and injected into it an element of dignity and effortless discipline. 'There might easily have been a disastrous drop in morale after Cotton left, 'writes Constance Babington Smith,[1] herself a fully-fledged member of 'Cotton's Circus' on the interpretation side. 'But Tuttle had the good sense to accept the flying club atmosphere, and not to try to regularise things all at once. He was quite prepared to overlook a pilot's blue suede shoes if that pilot was getting good photographs.' With the fall of France the unit had become vital to the country, and though for a long time Spitfires were necessarily in short supply, expansion was rapid. A move was made to the Thames Valley as Heston Airport became too small and too insecure.

Under Tuttle the unit continued with its high-speed high-altitude cover, and Cotton's interpreters with their stereoscopes and the Wild machines were in the centre of the scientific branch. Lemnos Hemming had succeeded in claiming the services of Claude Wavell, a mathematician who had worked with him on a pioneering air survey of Rio de Janeiro. The tendency with high-grade interpreters was specialisation, and this, sensibly, was encouraged. Michael Spender, for example, continued to specialise in German ships and harbours, Constance Babington Smith in German aircraft (and later rockets), and Claude Wavell became a specialist in German radio and Radar installations, a department that drew him much into contact with Dr R. V. Jones, Charles Frank, and Derek Garrard.

'Early in 1941,' writes Miss Babington Smith,[2] 'not long after Peter Riddell[3] had asked me to start an aircraft section, the Photographic Interpretation Unit [of the PRU] moved to a safer and more pleasant spot. Its new home was a large pseudo-Tudor mansion called Danesfield, a pretentious edifice of whitish-grey stone with castellated towers and fancy brick chimneys, which looks out southwards from a magnificent site between Marlow and Henley. When Danesfield became an Air Force station it had to be given an official title and it was named RAF Medmenham, after the little riverside village near by. From then on the name Medmenham—which in the eighteenth century had been linked with black magic because the first headquarters of the Hell Fire Club was at Medmenham Abbey—was identified with Photographic Intelligence. At Medmenham the Second Phase Section worked on day and night shifts in one of Danesfield's palatial halls. Outside the high west window was a mass of mauve wisteria, and the sweet, heavy scent drifted into Second Phase. . . . The distance that separated Medmenham from the [PRU] airfield at Benson was a gulf that was not very often crossed, which was a great pity. Whenever any of us did get over to meet the pilots, or when, during spells of bad weather, some of the pilots came over to Medmenham, it always left one feeling, Why can't this happen more often? Both our own work and the pilots' seemed to get a tremendous boost from it.'

That was not the only boost that PRU pilots were getting. They had learned the technique of 'dicing' (a good bit of RAF slang). When close-up photographic cover was urgently wanted of any place or object, the pilot would streak in low, very low indeed, and try to take the target as he shaved past it. Dicing was becoming so important that Tuttle had had the Spitfires painted pale pink, since that colour had been proved on trials to be better (low down) than Cotton's duck-egg green.

* * *

Jones, as one would expect, was extremely methodical, and liable to hang on to any clue. Having learned, in July, that a *Freya* on Cap de la Hague had been responsible, or partly so, for sinking HMS *Delight* he had asked PRU to cover Cap de la Hague and the village on it, Auderville. And he had asked Claude Wavell to bring him any pictures.

On November 22 Wavell turned up in London with prints which showed a couple of rather odd circles west of Auderville. He thought they might be 'cow pens'. But according to the Wild measurements, the 'pens' were only twenty feet in diameter, which did not look agricultural. They might be flak gun sites. Jones was very interested indeed. In the meantime he had established a technique for identifying *Knickebeins* by comparing the shadows of their big aerials as they rotated. When the Auderville pictures had come in, and while Jones was still talking with Wavell, Charles Frank put two similar pictures in his stereoscope, and, thinking of Jones's technique with *Knickebein* shadows, looked carefully at the 'cow pens'. It seemed fantastic, but he saw that the two pictures did not *quite* match up. A shadow inside one of the pens had fractionally widened. Exactly what time had elapsed between the two frames? Nine seconds. During those seconds, it seemed to him, some thin structure, perhaps a high and quite a wide structure, but thin, like a playing card standing on edge, had turned a bit on its central vertical axis. An aerial? It could be a Radar aerial. Had they found the villain that had been the end of *Delight*? Jones agreed with him; it looked like a breakthrough. PRU was asked to lay on dicing sorties over the pens west of Auderville.

There was a prolonged delay. Photo Reconnaissance was extended on anti-invasion assignments; and some of the senior officers running the war were very annoyed with Jones. It seemed that he was a prophet of doom. Now he was actually saying that the Hun had Radar! Even if he had, did it matter so much? Radar, after all (they thought as the Germans had) was defensive, did not kill people. But the Prime Minister was excited about it, and a high-powered committee under Joubert had been set up to get at the truth.

Then the first dicing sortie was a failure. The pilot photographed the next field to the 'cow pens', which took him right over an anti-aircraft gun. Jones was told that what he had taken to be enemy Radar was in fact a gun. But Jones sharply called for another photograph, this time on target. So, on February 22, 1941, another Spitfire pilot, Flying Officer W. K. Manifould, made the second attempt, and came back with a wonderful oblique.

One can imagine with what intense excitement, rather like Holmes and Watson first seeing the Hound of the Baskervilles, Jones, Frank, and Garrard examined Manifould's work of nerve and skill. So that was what a *Freya* looked like: beautiful! beautiful! What was more, that same day they picked up the *Freya* pulses from Auderville on a frequency of 120 mc/s. Garrard put them on a cathode-ray tube and broke them down into a perfectly intelligible signal. He noted that the signals were different from the *Seetakt* emissions that he had originally picked up on the cliffs of Dover.

Air Marshal Joubert de la Ferté was energetically pushing the official enquiry into the possibility of the Germans having Radar, and a meeting had been fixed in advance for February 23. When Dr Jones arrived at the meeting with his double proof (pictures and sound), the Air Marshal wondered if he was not being sprung—if Jones had not been piling up information in order to make a point. However the previous day's date on Manifould's oblique and Garrard's report proved that this was not so. The enquiry was ended; the question was answered.

Information now came in from Belgium and France. *Freyas* were much smaller than Chain Home towers, but turning aerials could not always be concealed. Also the British by radio and PRU searches were able to plot the *Freyas* round the length of the Kammhuber Line. A piece of ciné film was sent to England which showed Luftwaffe signallers working a *Freya* and tracking British aircraft. So much for *Freya,* but that there was another type of German Radar set, complementary, perhaps more important, the British knew.

'Ferret' Wellingtons of 109 Squadron in their radio-search sorties reported from the *Freya* sites different signals of shorter range on a frequency of 570 mc/s. One Wellington, on a coastal tour of Brittany on May 8, plotted nine of the 570 mc/s emissions. It was particularly maddening that nothing showed on any photographs of those places. Agents invited to list all known sorts of German Radar in the occupied territories spoke of FMG and *Freya.* The British did not know what FMG meant (actually the letters represented *Funk Messgerät*—radio measuring apparatus). Jones had learned that four FMGs were operating in Vienna, of all places, by the end of 1941. As Vienna seemed unstrategic, though emotionally important to Hitler and the Germans, Jones inferred that FMG was an integral part of German aerial defence, and had been produced in considerable numbers.

Two additional leads now came through to London, one from the United States, and the other from China. . . . Jones received a photograph taken by 'a well-wisher' in the American Embassy in Berlin. The photograph showed in the distance, above the tree-tops, a new German flak tower, and on the tower was a metal lattice aerial shaped like a saucer on edge. There was no foreground to the picture, and nothing to give scale to the aerial. But a few weeks later a Chinese scientist sent in a report. He had been walking near the Berlin Zoo and had seen the same flak tower from close-to. He described what he took to be a directional and ranging device for the guns in the tower; it was, he said, paraboloid in shape, and the diameter of the bowl was at least twenty feet.

His description puzzled the British. (He was, of course, describing one of the new Giant *Würzburg* sets that had been produced in response to Kammhuber's complaints about the original set.) If scores of 570 mc/s Radar installations existed round the coasts of German-held Europe, why were they invisible, if they were as big as the one near the Berlin Zoo? Could they be underground, or camouflaged in some special manner?

However, there were now more opportunities for photographic cover, and Charles Frank was looking at a new batch of aerial photographs when, once more, he thought he noticed something. It was on a medium-level photograph of a *Freya* station on the cliffs north of

Bruneval, between Le Havre and Etretat. Near the *Freya* station was the prominent lighthouse of Cap d'Antifer. Inland was a large farm complex round a square. That was where the signallers operating the station would be quartered. There were flak guns, well sited as usual. Between the big farm and Bruneval village, down below in its gulch, stood an isolated house. It would be officers' quarters, presumably, or some kind of headquarters. There were trees round the farm and inland, and trees round Bruneval. But the plateau surrounding the isolated house was bare, its turf eaten short by cattle. The paths or tracks from the *Freya* station to the farm, and from the farm to the house, showed up clearly on the photograph. But the equally well-used path from the station to the house was indirect. It led south to a small black dot, and from the dot it went up to the house. Frank admitted that the dot could be anything from a latrine to the entry of an underground dormitory. But there was a chance that it was the missing link, the Radar whose 570 mc/s signals the Ferrets had heard. There was only one thing to be done—call for dicing. Jones asked for low obliques of Bruneval 'through the regular channels', but whether by accident or design, he also let Claude Wavell know about it.

Jones had paid close attention to 'Cotton's Circus' and its development into the official photographic reconnaissance. He had seen, as Cotton and only a few others had, that when the country was on the defensive and when it went into the attack, photography with proper interpretation could tell more than a thousand agents. No part of PRU was without interest to him.

He first heard of Tony Hill when he was told that there was a PRU pilot who was immensely keen on taking low obliques (dicing), who had put in an immense amount of practice, but who always somehow foozled the camera shot, so that he just missed the vital object. Jones was interested. He made a point of meeting Hill, and he was, it seemed to Jones, a most likeable young flying officer, modest, shy, half-wild, dedicated. When they drank a pint of beer together in a Thames Valley pub Hill said that Jones ought to taste his father's beer. (His father, Colonel Hill, ran a brewery in Hertfordshire. He had brewed the ale for the 1937 Coronation.) Jones did not taste that beer until the war was over, and then Tony Hill was not there. When they discussed dicing, Tony explained that he was by nature 'a bit slow', and he just did not seem to be able to manage it. So the scientist and the pilot took the problem to bits and put it together again. The timing

The ground reconnaissance: Rémy, head of the Channel Coast resistance network; pictured while taking oath at trial of spy who destroyed his network. (*George Millar*)

The air reconnaissance: Tony Hill's 'dicing' oblique shows the small (10 ft diameter) *Würzburg* and the Lone House. (*Imperial War Museum*)

*Above* Bruneval, during an Anglo-French commemoration service. In the background, the cliff on which the *Würzburg* stood (to left of building). In the middle-ground, the gulley from which the German machine-gun fired up the hill. (*C. W. H. Cox*)
*Below* Bruneval: the beach seen from the *Würzburg* site. In the foreground, the hill down which the withdrawal had to be made; on the left, the Bruneval road. (*C. W. H. Cox*)

was, in fact, very difficult, because the camera was not forward-facing, and therefore was out of relation with the pilot's vision and his think-box. The camera was behind the pilot, and pointing broadside and a little astern. He therefore had to dive on the target and watch it disappear under that shark-like Spitfire wing. Then it flashed up on the other side. Just where he pressed the button depended on his personal reactions, and his nerve, and a few other factors. He simply had to learn by trial and error. When they had talked it all out, and reduced it virtually to a habit, like golf or tennis, Hill practised a lot more, got the knack of it, and went on to take some of the best photographs of the whole war.

At this stage in the hunt there occurred one of those 'rare but precious visits' of PRU pilots from Benson in Oxfordshire to their Intelligence Wing at the wisteria-draped Danesfield, fifteen miles away. Gordon Hughes drove over with Tony Hill. Hughes made his way to Claude Wavell's office, where Wavell at once showed him his newest gadget. He had called it, rather grandly, the Altazimeter, and he had designed and made it himself. Its function was to work out rapidly and accurately the actual heights of objects seen on aerial pictures. Height, he explained to Hughes, equalled shadow-length times the tangent of the sun's altitude. It followed that the only data needed were latitude, the scale of the photograph(s), the orientation, and the date.[4] Simple, eh? Merely an application of the principles of spherical trig. That was the kind of thing that Gordon Hughes came to Danesfield to hear. He found it both elevating and restful. But now Wavell changed the subject.

'Look here!' He took from his desk two photographs of a length of cliffy coastline and put them in his stereoscope. As Hughes looked Wavell explained. It was felt that the speck where that path to the isolated house suddenly changed course might be the paraboloidal installation (it sounded vaguely Chinese) that the scientists were all so keen to find.

Hughes suddenly remembered that Tony Hill was waiting for him below. He would really have to go.

'Hill!' Wavell said. 'Ask him to come and have a squint at this.'

When Hill was sitting opposite him Wavell began to talk about German RDF. It was different from the English variety, and its origins were as old or older. He laid another pair of pictures under the stereoscope, and showed Hill the one millimetre variation in shadow that had put Charles Frank on the track of the *Freya*. That piece of inspired deduction, he said, had led to *this*. . . . He produced the Manifould oblique. Until that bit of dicing had been brought off people had refused to believe that the Germans *had* RDF. The oblique was proof, and it showed the equipment to be very sophisticated. It wasn't funny. Bomber losses over Germany were getting worse, and the enemy RDF plainly had much to do with it, and especially, it was thought, the as yet undiscovered set. The one that worked on 570 megacycles. He had changed the photographs again, putting the first two back on top. Was the dot the paraboloidal installation? . . . Hill spoke.

'Where exactly is this place, Bruneval?'

Wavell told him.

'I'll get you your answer tomorrow.'

Next morning Wavell saw on the operations' board:

Etretat—Hill

and he was worried. What had he done? The flak would be hot. He telephoned Benson. Would Mr Hill be good enough to call him at RAF Medmenham as soon as he got back?

His heart turned over when the telephone rang.

'Tony Hill here. You were right. It must be a paraboloidal whatnot and the Jerries were round it like flies.'

'You saw it?'

'Clearly. It's like an electric bowl fire and about ten foot across.' Hill then apologised. His camera had failed. There were no pictures. 'But don't worry. I'll have another go tomorrow.'

There was a rule that no pilot might fly the same dicing sortie two days running. The risks were great the first day, and on the second, with an alerted defence, they could be expected to be many times greater. Hill was determined to photograph the 'bowl fire', and nothing, nothing was going to stop him. He had no authorisation, and as luck would have it Jones's official request had gone to another squadron (Army Co-operation), three of whose aircraft were warming up, also at Benson, as Tony Hill got into his Spitfire. A senior member of the ground people came and challenged his right to go. Hill sent a message across the aerodrome to the other pilots, saying that the Bruneval job was his, and that if he saw them within twenty miles of the target he would shoot them down (he did not specify what with). In any event, he swooped again over Bruneval, shaving the edge of the cliff and the lone, ugly house, the scream of speed in his head, the surge of the Merlin at his knees. And the low oblique he brought back to Benson was just what was wanted. He had brought the *Würzburg* to life.

Hill's next great pictures were of two Giant *Würzburgs* on the Dutch island of Walcheren. Jones's office had news of a German night fighter station there and high-level photographs indicated one *Freya* and two Berlin-Zoo type *Würzburgs*. Hill swept down in a lower than low curve along the Dutch coast and his sideways-aimed camera took from point blank range first one Giant and then the other. They were facing different ways as he passed, so the German apparatus was taken from two angles. He also caught one of the Luftwaffe Signals crew of the second *Würzburg*. He was just climbing to the cabin, and 'froze' on the metal ladder as the Englander passed.

That was in May, 1942. In October of the same year Hill diced an aircraft factory at Le Creusot. He was not satisfied with his first flight, so again, as at Bruneval, repeated it the following day, and was shot down.

'We had become very good friends,' Jones says.[5] 'And I think I felt Tony's death at Le Creusot on October 21, 1942, more than any other during the war. He was very much my idea of what a schoolboy's hero should be. As soon as I heard that he was down we organised a rescue operation. But he was too badly injured to survive.'

## NOTES TO CHAPTER 15

1. Babington Smith, *Evidence in Camera*, p. 68.
2. Babington Smith, *op. cit.*, p. 107.
3. Sidney Cotton in February, 1940 had noted that Bomber Command had appointed 'an imaginative and knowledgeable regular officer named Peter Riddell' to organise a photographic and interpretation centre at High Wycombe.
4. Babington Smith, *op. cit.*, p. 167.
5. Jones, letter to the author.

# 16

# *Coupe Singapour*

### *February, 1942*

February began badly for the patriot Rémy. His impending Lysander flight from France to England was so frequently postponed that it seemed for ever unlikely. Yet it hung over his life and disrupted his other activities, and at a time of more than usual difficulty for the Confrérie Notre Dame. True, he had been cheered by Pol's luck with the Bruneval report, and it had seemed too good to be true when the irrepressible Bob, on the 9th, transmitted both messages to London.

But on the evening of the 12th he bought a *Paris Soir* and read with disgust the main news story. *Scharnhorst*, *Gneisenau*, and *Prinz Eugen* were said to have left Brest the night before and to have steamed up-Channel for Germany without being challenged by the Royal Navy or the Royal Air Force. Rémy was enraged. The escape seemed to set at naught the achievements and sacrifices of his CND agents, particularly those of Hilarion and L'Hermite. L'Hermite (Bernard Anquetil) had been killed by a firing squad the previous October at Mont Valérien. He had refused to utter a single word that might have incriminated CND or any of its host of fringe helpers. Had he died in vain? It truly seemed so. Had the English paid absolutely no attention to accurate information gleaned by Hilarion from the radish patches of the Arsenal garden? It would seem so. Had Pol's reports of sudden moves of German fighters to the Channel airfields been regarded as so much eyewash?

On the other hand, Rémy was ruefully aware that the English, thanks to those pestilential Nippons and to the Germans with their submarines and their Afrika Korps, were all too busy elsewhere. Everything, but everything, seemed to be going wrong with the right side. . . . And the blasted French newspapers, apparently happy to take their lead from Dr Göbbels' Propaganda Ministry, trumpeted out each Axis advantage, hailing every skirmish as a major victory. As to their panegyrics on the escape of the three German ships from Brest—anyone would have thought, as Rémy pointed out, that Nelson had lost the Battle of Trafalgar. He argued with any of his friends who had the heart to discuss so sore a subject that the German warships had been blockaded in Brest for close on a year. And now, instead of surging into the Atlantic to prey on Allied shipping and shoot it out with the Royal Navy, they were scampering east to the comparative safety of their home ports. Comparative only: the British had shown and were showing, that they could bomb all Germany. He produced another excuse. The Germans had picked very thick, typical Channel weather. The English spy plane of Coastal Command that always shadowed the three ships, for so long stationary, had caught just one glimpse of them sneaking out through the Goulet de Brest, and his wireless had chosen that very moment to break down. How unlucky could one be? All the same, even he had to admit that the Germans had conceived a dangerous operation, and had carried it through successfully. They had sent the three ships up-Channel under a tremendous fighter cover. Where then was the RAF, which had been victorious in the Battle of Britain? Where was the Royal Navy? Where the heavy coastal artillery, the submarines, the minefields, the swift motor torpedo-boats the English were said to possess?

He would have been surprised (but perhaps not relieved) to know that a small section of English opinion, an increasingly important section, was not displeased, since the German break-out emphasised the importance of science in the coming phases of the war. The general public, the Prime Minister and his colleagues, the press were even more dismayed than Rémy; *The Times* described it as 'the most mortifying episode in our naval history since the Dutch got inside the Thames in the seventeenth century'. What was not understood, how could it be—was that with the escape of the three ships the Germans had called an end to the Radar truce, since they had achieved the escape through temporarily blotting out British Radar. For the Radar war, a jamming war, a war of electronics, that was to ensue, the British and their new great ally were immeasurably the stronger and the more ingenious side. Again, as Rémy rightly sensed, in moving the three ships as he had done into the dubious security of North Sea ports, Hitler had given the first lair-like indication of the war. He was drawing into a defensive ring. And the reason was not far to seek. Air power. It was no longer good tactics to send capital ships marauding. They were too easily found and destroyed—from the air. The Luftwaffe, against expectations, had just failed to get on top of the Royal Air Force. It was the Battle of Britain result, and Bomber Command's aggression, that made him call the three warships home rather than send them out into the Atlantic to dig a huge hole in England's supplies.

The German anti-Radar moves (one cannot dignify them with the name of campaign) during the Battle of Britain had been mishandled in a fairly logical way. While Chain Home stations were attacked Martini's people on the French coast listened. As Signals reported that bombed stations declared destroyed continued to transmit, it came to be thought at OKL that the important 'innards' of each British station must be underground, in bomb-proof bunkers (which was not the case). Göring having decided, accordingly, to discontinue such attacks, any further bombing that came was spasmodic, and the defence's remarkable repair services easily coped.

Meanwhile, however, Martini had set up, on the Cherbourg Peninsula and near Calais, cross-Channel jamming transmitters. After their first shock at realising the extent, and depth, of the British Radar coverage, Luftwaffe Signals had understood that the British Herzian walls were by no means invulnerable. Many people in Britain held the same opinion. When A. P. Rowe took over the running of Bawdesey from Watson-Watt in August 1938 he asked Dr E. C. Williams to try jamming the station. A diathermy set was accordingly fitted in a Sunderland flying boat. When this heavy machine wallowed into Bawdsey's Radar twilight the Radar screens were covered with pretty dancing lights. Coloured filters helped the (then inexperienced) operators to distinguish the trace. And in subsequent experiments Professor T. R. Merton had developed 'the long afterglow' whereby the aircraft trace remained while the dancing lights died.

In the early days of September 1940 the first German jamming transmitters began their effort. This ran at full bore on September 11. By using coloured slides and waiting for the afterglow the operators, many of them now WAAFs and very much on top of their job, kept going, and the Chain Home cover of the South Coast was never completely at a loss. That it was seriously interfered with was indicated by the day's results, which were unusually bad. Fighter Command flew 678 sorties, losing twenty-nine aircraft with seventeen pilots killed and six wounded. The Luftwaffe lost only twenty-five aircraft. It was most unusual. Two days later, when the Germans attacked London intermittently by day and fiercely at night, there was an attempt to jam, and three Chain Home stations, Dover, Rye, and Canewdon were in some difficulty. The general impression, on *both* sides, was that the German jammers were not powerful enough. The British felt that airborne jammers would have been more effective. In that stage of the war the German jamming, like their direct attacks on Radar stations, had been too haphazard. Nor had the importance of Radar been appreciated by Germany's leaders. At his conference in Berlin on September 13, Hitler referred to the much worse than usual showing of the RAF on September 11. But he did not in any way link it with Luftwaffe Signals' jamming. He merely said that he thought the British defences were at last cracking. One gets the impression that he thought in terms of a kind of poster art—Nordic heroic youth battering at a more collapsible form of youth.

This emotional slant came into the move of the *Scharnhorst*, *Gneisenau*, and *Prinz Eugen*, which was, strategically, a bad move. At Brest the German ships were tying down important British forces and were ready to exploit any obvious British weakness in the Atlantic battle. However, he could not bear to think of them being bombed in a foreign land, and he had visions of a naval power centre in the North Sea, and even of the Russian Navy then, as now, strong on paper, offering a threat. But there was nothing haphazard or untechnical about his orders for the move. *Complete* Luftwaffe cover had to be given. Göring, nervous about the responsibility, handed it over to Field Marshal Milch. Milch went into it with Martini, whom he disliked and despised. Dr Hans Plendl, the inventor of two of Jones's headaches, shortly now to be given a senior post in Germany's Radar defences organisation, was also called in to advise. A programme of analysis, location (with special narrow-angle direction finders suggested by Plendl, and with the help of Luftwaffe reconnaissance), and frequency-testing of the British Radar service on the South Coast was rapidly carried through. The Germans were now in close contact with their main ally, the Japanese, and were beginning to understand that ships would not function without air superiority. They now agreed that air superiority had not been achieved over England. But, by decree of the Führer, it *must* be achieved over those three ships for two days. And it was so achieved.[1] It was found that the British Radar cover was limited in frequency variation, and could easily be jammed. Proper equipment was set up in two centres directly connected by landline and therefore controllable from a central headquarters. These were strategically placed near Cherbourg, to cover the wide part of the Channel, and at Calais, for the vital narrows. A co-ordinated programme was worked out with the Luftwaffe fighters. So, while the fighter screen was gathering in the North of France the jammers from the two centres began gradually getting the British operators used to spasmodic interference. Then, during the escape, when the three ships were already well up-Channel from Brest, the jammers opened up, and for a while in that part of England there was no Radar cover.

A board of enquiry was set up in England to ascertain how the German ships had managed to move almost unhindered so close to the South Coast. The true results of the enquiry were naturally unpublicised. These were decisions to broaden the frequencies of many British Radar services, making them less vulnerable to jamming, and, far more important, to become aggressors in the Radar war. Up to this point Radar had helped to save the country and had proved itself in a defensive role: *Knickebein* had been jammed, and *X-Gerät*, and *Y-Gerät*. It was now time to attack the Radar systems on German and German-held territory.[2] Cockburn and his helpers at Swanage had a means ready to dismay the German *Freya* system. As for the *Würzburg*, well, that would be tackled after the Bruneval Raid.

* * *

There was some pallid relief for Rémy following the break-out from Brest since the British bombing of that town ceased and Hilarion was no longer in danger of death and worse than death. Rémy was still under orders from London to remain in Paris until Operation Julie, that celebrated unbelievable flight across the Channel. He objected to being thus confined, and the news from the Far East made him miserable.

On the 15th Singapore capitulated to the Japanese. Where, oh where, would it all end? The 'island fortress' had held out for only two weeks after the day when all the British forces left in the Malay Peninsula withdrew within its perimeters. Churchill had at once admitted the disaster to be one of the worst in British history. Göbbels was, for once, in enthusiastic accord, and as usual he got a lot of coverage in the German-influenced newspapers of Paris.

Infuriated by all that he had read, Rémy found himself at lunchtime in one of the premier restaurants surrounded by German officers who seemed to him 'more arrogant than ever'. And, culminating stroke, when he was handed the vast menu he was outraged to read, 'fully visible in the middle of the fish and shellfish dishes' the words *Coupe Singapour*. He made up his mind to have a reckoning with the restaurateur after a liberation that that morning seemed horribly distant, even uncertain.

## NOTES TO CHAPTER 16

1. Galland, *The First and the Last*, pp, 140–67. Summing up the operation, Galland attributes much of its success to 'a clever trick' of Martini. He adds: 'Unfortunately the German command did not draw the necessary conclusions from this victory in the Radar war. . . . The British learned from their defeat and developed Radar interference to a perfection which, in the later bombing war, became fatal for the Reich,' p. 165.
2. Price, *Instruments of Darkness*, p. 88.

# 17
# Henry

*February, 1942*

When the Bruneval low oblique reached his office in the Air Ministry Jones said to his deputy, 'Charles, we could get in there; there's a beach only a few hundred yards from the objective.' Almost the next person he happened to meet was W. B. Lewis, Deputy Superintendent of TRE. Lewis said that any proposal to raid the cliff top for samples would have his support. Jones then took his suggestion to the Assistant Chief of Air Staff, and he mentioned it to Lord Cherwell, which was as good as telling Churchill. It would seem that the firm proposal for the Bruneval Raid went from the Air Staff (who were interested in the end result) to Combined Operations (who would have to do the raiding). At a slightly later stage Tizard, independently, had the same idea.

Derek Garrard asked Jones if he might volunteer to go on the raid as one of the demolition squad. Jones put this forward to other members of the Air Staff, but he was soon told that Portal had 'very reasonably put an embargo on Garrard, myself, or any member of my staff'. From then on he watched, as an informed outsider, the progress of 'C' Company's training. What did directly concern him was the *Würzburg* itself, and that meant a contact with Flight-Sergeant Cox and Lieutenant Vernon. The date of the raid was drawing very near when Cox and Vernon were granted two days 'compassionate' leave. Each was told in confidence to report in the afternoon of the second day to the duty officer at the Air Ministry.

'I went home for one precious night to Wisbech,' Cox says.[1] 'When I got to the Air Ministry the next day Lieutenant Vernon was there, in the waiting room. I thought it peculiar, him not being an airman. We were taken to an office. Three men sat behind the desk, two Englishmen in civilian clothes and a Frenchman in British Army uniform, battle dress. He said little. The Englishman in the middle, powerfully built, sure of himself, a good bit younger than me, did the talking, while the third man spoke occasionally when what you would call the technical side of matters was under discussion.

'There was a lot of talk to begin with about what would happen if we were taken by the Jerries. We were only to tell them, of course, the standard things, name, rank, and number. But we must make it clear, if caught on the job, that we were simply a demolition squad out to do mischief to a valuable bit of enemy equipment. We would both come in for special questioning, since I was the only airman in the parachuting party, and Vernon was the only engineer officer. We discussed alibis. The tall man said the Germans often planted an "English" fellow-prisoner in the cell of a newly-captured man. He would be an expert at getting information, and there could be hidden mikes. We were warned too against the kindness-and-generosity treatment: they might put you in a comfortable room with soft music, a box of Coronas, and a bottle of whisky or brandy. . . . I told the Intelligence people I could stand up to any amount of *that* type of interrogation.

'We were given advice on how to escape capture if things went badly wrong with the raid. With the French people in the locality we would be among friends. Granted half a chance, any of the farmers or villagers would hide us and risk their lives, and more than their own lives for us, just as they were doing for all our boys shot down over France. It might well be, the big man said, that even if things went wrong, we would be smuggled, Dennis and me, fairly quickly back to England, either by boat or in a small pick-up aircraft. But in case we were completely on our own over there we were given French money, maps printed on fine silk, and collar studs with miniature compasses hidden in the bases. We had to memorise three addresses, two in France and one in Switzerland. If we got to any of the three we had a code password. The people in the houses would do the rest. We would simply be packages in their care.

'Most of the rest of the interview was technical stuff concerning the German RDF set, what was wanted from it, and why. Both the Englishmen evidently knew more about Radar than we did. Without raising his voice or saying anything dramatic, the main spokesman had made us feel that our job was something really worth doing, and that we were lucky to find ourselves doing it.

'It did not take long. It was very matter-of-fact and reassuring. A WAAF driver took us in a staff Humber to Waterloo Station. When we left the train at Salisbury, another WAAF recognised us, guided us to a similar Humber, and drove us to Tilshead. We were beginning to feel quite important people; but we were soon cut down to size, the pair of us, because our lot were still doing those damned exercises with the Navy, and it was too near mid-winter for bathing. (Supposing either Dennis Vernon or I had caught pneumonia, and hadn't been able to go on the raid?)'

Jones's companion at the interview was a Radar specialist from TRE at Swanage, D. H. Priest. Dangerous though it was judged to be in view of the known 'thoroughness' of German interrogation methods, it had been decided to send one expert. Priest had been chosen, and had been given a temporary Flight-Lieutenant's commission and a cover story for the occasion. He was to arrive with the landing craft, and his code name was NOAH. It was envisaged that *if* the fight went easily for the parachutists up above on the headland as well as down by the beach, and *if* the landing craft made the beach in good time, and *if* armoured German units were slower than seemed likely in arriving at the scene, then Priest might have time to hurry up the cliff path, make an assessment of the paraboloid, and supervise its dismemberment. It was a tall order. But they had to think of every eventuality.

To John Frost and 'C' Company as a whole, the successful breakout of *Sharnhorst*, *Gneisenau*, and *Prinz Eugen*, was upsetting. It appeared to show a certain British lack of control in the Channel area, and did not augur well for their own evacuation. And the marine side of training stubbornly refused to go properly; indeed, it did not go at all.

'Security' would not sanction a move of the parachute unit to new quarters nearer the South Coast. Accordingly, day after short February day, they drove south in Bedford trucks from Tilshead, did their watery exercise and returned in the dark. The Dorset coast which, particularly round Lulworth, has cliffs resembling those at Bruneval, seemed to be favoured by the Navy. It is an exposed coast, and the weather was obstinately foul. What had been planned as the final rehearsal was one of the worst of a series of disappointments. The parachutists were to leave their transport on a flat stretch of land near the sea. The Whitleys would there parachute the weapon containers and the folding trolley (a light-weight but very strong two-wheeled trolley with a good capacity for stolen mechanical parts, specially designed and made to drop on its own parachute). And after attacking an imaginary objective, the raiding party would consolidate on the beach and call in the landing craft by wireless and by radio beacon. As it happened, the landing craft waited off the wrong beach, the Whitleys dropped the containers in the wrong place, and the parachutists got trapped in a defensive minefield, and were lucky to get out of it without casualties.

When only forty-eight hours remained before the first possible departure on the raid itself, the naval authority at Portsmouth insisted on yet another night exercise. This had to be postponed for twenty-four hours because of heavy weather. It was then held on Sunday, February 22, in Southampton Water; there was still far too much wind, outside in the English Channel.

This time 'C' Company with all its gear and appendages made a 'withdrawal' in creditable silence and good order. Four men pulled and pushed the two-wheeled trolley, carrying a sizeable boulder. The timing was correct. Contact had (for once) been establishedwith the Navy on both No. 18 wireless sets. The landing craft were all there—but a long way offshore. They were ordered to wade out to the boats. It seemed pointless to risk their weapons and the trolley; these were left under guard at the water's edge. For a considerable distance the piercingly cold sea was only three feet deep. Then it crept up to their thighs, then shoaled again. Cox had to make a dash for the boats. His legs were stiffening with cramp. When the hundred-and-sixteen cursing men were aboard, the landing craft went full astern—and did not budge. The officers jumped overboard, followed by the men. Everybody pushed and heaved. Still the boats, aground on a falling tide, would not move. They waded ashore, facing the tedious, too familiar, drive in soaking clothes and draughty trucks, back to the dark huts of Tilshead. The next four days, Monday to Thursday, were the only possible days for the operation in February, when the moon was full and the landing craft could approach Bruneval on a rising tide.

Orders had now been issued. All the officers had typewritten copies headed:

SECRET

NOT TO BE TAKEN IN AIRCRAFT

OPERATIONAL ORDER 'BITING'

By

Major J. D. Frost, Commanding

'C' Coy, 2 Parachute Bn.

*Topography*—Scene of operation to be explained on model.

In the cellars of Danesfield, below the 'palatial halls' where Constance Babington Smith and Claude Wavell and others worked at interpretation for photographic reconnaissance, there was a model-making workshop. The Bruneval model had been accurately put together from blown-up air photographs and the biggest-scale French maps. The *Würzburg* was there, the ugly house, trees, fences, gates, German pillboxes, all to scale. Since it was to be an operation by moonlight and over complicated and rugged terrain, the lie of the land had to be familiar to every man. Few people can 'see' a piece of country by looking at a map. A model is quite different. In this case, too, it cut out the need for describing places by their foreign names.

In the Orders the German defences round the beach were described as BEACH FORT, REDOUBT, and GUARD ROOM. The house, locally called the château, was LONE HOUSE, and the big farmyard of Le Presbytère was RECTANGLE. The *Würzburg* was HENRY.

The stated *Objects* of the raid were: 'To capture various parts of HENRY and bring them down to the boats. To capture prisoners who have been in charge of HENRY. And to obtain all possible information about HENRY, and any documents referring to him which may be in LONE HOUSE.'

Frost's doubts as to the merits of the original Divisional plan had not induced the planners to change it. His force was divided into three parties of unequal size, each with its own limited objectives. The parties were to drop at five-minute intervals, the first sticks going down at fifteen minutes past midnight.

The First Party was named NELSON. Because it had the longest journey to its objective it was to drop first, and to move silently and swiftly down to the beach. When Frost gave the attacking signal up above, this party was to storm, take, and hold REDOUBT, BEACH FORT, and GUARD ROOM, thus securing the whole force's line of retreat. This was the biggest party of the three, and its task was probably the most difficult. It consisted of three light assault sections commanded by Lieutenant E. C. B. ('Junior') Charteris of the King's Own Scottish Borderers, and the heavy section under Captain John Ross, Black Watch. In his section Ross had some sappers with anti-tank mines (for the Bruneval road to the beach) and mine detectors and guide lines. As soon as the beach was taken he was to mark a route through the minefield, if any, establish a check point, and see that the signallers contacted the Royal Navy. It was this part of the plan, Frost believed, where most could go wrong. There was no saying how many Germans would be in the beach defences after midnight, assuming that surprise was achieved; and if strongly and resolutely manned they could be all but impregnable. Charteris and Ross were both officers of character and determination to whom the men were devoted. Charteris's sections, mainly made up of Seaforth Highlanders, had tremendous offensive potentiality.

The Second Party had to split into three, and therefore had three names. DRAKE, under Lieutenant Peter Naumoff, was to move towards RECTANGLE and take up a position south of it to ward off enemy threats to LONE HOUSE and HENRY. HARDY, a small group under Frost, would surround LONE HOUSE, while JELLICOE, an assault section under Lieutenant Peter Young, would surround HENRY. The Royal Engineers' party, led by Vernon and accompanied by Cox, would move to HENRY when Young's party had cleared it of the enemy, and Young would then protect them while they worked.

The Third Party, RODNEY, thirty men under Lieutenant John Timothy, Royal West Kent Regiment, would drop last, and would act as landward screen, reserve, and, finally, rearguard.

As to timing: everybody was to get into his attacking position as quickly and silently as possible, and when this had been effected Frost, from the steps of LONE HOUSE, would give the signal to attack, four blasts on his whistle.

They all agreed that the Dropping Zone, due east of a track running north and south on the model, seemed a good one, and so did the Forming Up Point by a line of trees nearby. They also agreed that it would be a cheerful thing to see NOAH come up the hill from the beach. That would mean that the boats had actually reached the place, and that the situation at two key points, the beach and HENRY, was under some kind of control. There were surprisingly few questions. One man, when looking at Tony Hill's oblique, asked how surprise could be effected if the RAF went flashing cameras in the Germans' faces. 'Oh telephoto lens, you know. Taken from miles away,' Frost replied. Another asked if they would drop with blackened faces. 'Certainly not,' was the answer. 'The main thing on this party is to avoid confusion. I'll have no black faces. I want to be able to recognise you in the moonlight.'[2] And a third man asked, 'When you get to the front door of LONE HOUSE, sir, and you blow your whistle, what do you do if the door's locked?'

'Ring the bell.'

* * *

On the Monday morning 'C' Company went through their normal routine. At ten o'clock they checked and cleaned all weapons and packed the containers, which were to leave for Thruxton Aerodrome at two. Ross had contrived, without breaking security with the Glider Pilot Regiment, to get the men a particularly good midday meal, and after it they were urged to have a siesta until tea, at five o'clock. And at tea-time a message came through from Division. 'Owing to adverse weather conditions' there would have to be a twenty-four hour postponement.

Each morning they repacked the containers, and each evening they unpacked them. The weapons were cleaned, recleaned, and cleaned again. Tuesday, Wednesday, and Thursday were exact repeats of Monday; and Thursday had always been named as the very last possible day in February for the raid.

'We are all thoroughly miserable,' Frost wrote in his diary. 'Each morning we brace ourselves for the venture, and each night, after a further postponement, we have time to think of all the things that can go wrong, and to reflect that if we don't go on Thursday we shall have to wait for a whole month to pass before conditions *may* be suitable. After all, the weather in the English Channel in February and March is not inclined to be "suitable".'

Friday morning was bright and frosty. The wind seemed to have dropped. The clouds had gone. Frost expected a message from Divisional Headquarters instructing him to send everyone on leave. But the message was that the other arms involved had agreed to see what weather one more night would bring.

Stand-by again. For the fifth day running they went through the now tedious routine, breakfast, tidying up, containers packed. They all felt listless except, it seemed to Frost, Sergeant-Major Strachan, who was in high good humour and said he was sure they were going to have some fun at last.

At tea-time General Browning, immaculate as usual, arrived to wish them all luck. The raid was on.

*Above* One of the landing craft bringing back the Bruneval raiders. (*Imperial War Museum*)
*Below* Back at Portsmouth, the motor gunboats are secured alongside the *Prins Albert*. (*Imperial War Museum*)

*Above* Wing-Commander Charles Pickard, who led the flight of Whitleys which dropped the paratroops, examines a trophy of the Raid. (*Imperial War Museum*)
*Below* One of the German prisoners from Bruneval is handed over to The Military. (*Imperial War Museum*)

As the Royal Navy's participation called for an earlier start from England, Admiral Sir William James, Commander-in-Chief Portsmouth, had signalled that morning: 'Carry out Operation BITING tonight 27 February.'[3]

In the afternoon *Prins Albert*, carrying, apart from her regular complement, thirty-two Welsh commandos, officers and men of the Royal Fusiliers and the South Wales Borderers, slipped out through the boom defences escorted by five motor gun boats and two destroyers. Each MGB carried a crew of sixteen, and the 2,700 horse power of its three Hall-Scott petrol engines gave it a speed of twenty-seven knots. The armament was two two-pounders, two half-inch twins, one Oerlikon, and four depth charges. Long after dark, at 9.52, *Prins Albert* lowered six landing craft into the sea. Each, in addition to its regular crew, carried four soldiers who were there to give extra fire-power with their Bren guns. The Welshmen had blackened their faces and were in the highest of high spirits.

*Prins Albert* turned and headed home.

## NOTES TO CHAPTER 17

1. Cox's written account.
2. Frost, discussions with the author.
3. Price, *Instruments of Darkness*, p. 82.

# 18

# Aircraft that Pass in the Night

*February, 1942*

That Friday morning Rémy awoke at Saint-Saëns, north of Rouen and not far inland from Bruneval. He had endured a wretched night, sharing a bed with a grumbling companion also due to go to England by that unconventional means of transport, the Westland Lysander. Julitte, like Rémy, had been put off many times. He had *quite* lost hope, and every five minutes or so, he said so.

Then Rémy, a strong man, who at that moment was coughing with the heaviest kind of cold in the head accompanied by fever, was not the sort of bedfellow that even a cheerful soul would choose. But the café of the Guardian Angel (Marcel Legardien), who was once more sheltering them, could only offer one spare bedroom which held one bed.

Cautiously parting the curtains, Rémy peered out at the weather. A leaden sky. Cloud base, he thought, as low as six hundred metres. Bob had already left his bed on the floor and gone to gauge conditions at the landing ground.

'Not more than twenty centimetres of snow anywhere,' Bob announced on his return. 'It's hard snow, perfectly good for Lysander work.' (He had done a course in England on the organisation of 'reception committees'.)[1]

Julitte, Bob, and Rémy, had had an enormous slice of luck the day before. They left Paris by train, from the Gare St Lazare. Arrived at Rouen, they looked for the bus to Saint-Saëns. As usual in those days, the bus was packed with travellers, and the roof was smothered in an untidy mass of luggage and bicycles. Rémy, who had much natural authority, persuaded the driver to find room on the roof for their two pieces of luggage. One of these was Bob's English transmitting set in its standard brown suitcase, and the other was a bulky valise filled with CND documents on their way to England—a risky passage indeed! Having seen to their luggage, the three men with difficulty thrust themselves into the interior of the bus. None of them was too pleased, at the end of these victorious struggles, to see Léon (whom they all knew well) rushing across the square waving his arms. They had caught the bus believing the Guardian Angel's tyres to be completely worn out. But now a breathless Léon assured them that the Guardian Angel had been able to find new tyres on the black market, and that he was waiting for them with his van outside his brother's café. Rémy had to give the furious bus driver a large tip to get down the cases he had had such difficulty in stowing. They walked to the café near the Palais de Justice, finding their luggage heavy, but delighted after all that the Angel had come for them.

On their way to Saint-Saëns they passed the bus from Rouen station which had been stopped by a patrol of the Feldgendarmerie. The passengers were lined up at the roadside while every bit of luggage was dragged off the roof and searched . . . The risks that Rémy and his people thought nothing of running were the more terrible because sometimes they would have been avoidable.

On the Friday morning Bob made early contact with London and was told to keep listening watch.

At five in the afternoon London had a message for him. The Lysander flight, Operation Julie, was on.

But in the Personal Messages following the BBC's French Service news at 7.30 p.m. Operation Julie was postponed till the following night, and even the imperturbable Guardian Angel, rising to produce a bottle of his miraculous Calvados, exploded, 'What's the matter with them now, ces bougres-là? It's a perfectly gorgeous night.' Julitte did not actually say anything, but his expression—wrinkles of bitter laughter round the eyes, a venomous droop to his mouth—said plainer than words, 'Didn't I say so? That Lysander will *never* come.' Tasting the Calvados, his aching eyes shut, Rémy forced himself to remain calm, good humoured. This proposed flight to London had always seemed to him a waste of time. He was only going there because Passy, the increasingly powerful Passy, insisted on seeing him. So much time had been wasted when he could have been elsewhere in the CND area, looking after his own people, wasted in hundreds of hours of waiting in those dangerous corners, with far too many radio 'vacations'.

It really was too maddening—particularly when one had a cough, a temperature, sore sinuses, watering eyes, in fact a hellish cold in the head. The Guardian Angel produced a pack of cards. Oh well, it would pass the time. . . .

At 9.15 p.m. the Angel laid down his cards and said, 'Shall we hear the news?' Julitte quickly said, 'We've heard it.' Even Rémy shrugged, and did not feel like listening. But the Angel had firmly switched on, and the French voice from London worked through the familiar day's news to . . . 'Veuillez maintenant écouter quelques Messages Personnels . . .' Julitte laughed bitterly, and Rémy was distressed to find himself echoing the laughter. But suddenly the silence in the smoke-filled room where there had been too much time to pass was complete; and there was no time to waste. Their message had come through. Operation Julie was ON. And the full moon was early that night. And the rendez-vous had been fixed for between ten and eleven (German time). They dashed upstairs to dress, to grab their luggage, overcoats, gloves. Downstairs again, there was just time to kiss Madame Legardien in front of the round-eyed maid, and they were outside in the deserted snowy street of Saint-Saëns. Bob had hurried ahead with Léon to set up his triangle of lights. The sky was full of stars.

Now all three were heavy-laden, for they had to carry Rémy's trunk-full of papers for London together with the biggest-scale maps covering the whole of France, that Passy had ordered. The mass of paper had been divided between three sacks. Rémy carried his sack on his right shoulder. In his left hand was the equally compromising, and by no means light, suitcase. In order to shield his appalling cold against the night air of February he had covered his chest and his back with thick pads of thermogenised and medicated cotton wool. This ferocious breastplate was held in place by a woollen vest covered by a lumber-jack's shirt which, in its turn was covered by a massive jersey in white wool knitted by Madame Edith. Over this he wore yet another thickness of wool. He had buttoned his tweed jacket with the greatest difficulty. Enormously thick stockings came up to his knickerbockers. A Balaclava helmet covered his Basque beret, and he sported fur-lined gauntlets and a fur-lined leather coat, a Canadienne, then the most popular article of clothing in France. Never a sylph, Rémy admits that that moonlit night he must have looked like a perambulating barrel. The Guardian Angel led uphill through snow, and the landing ground was at least one kilometre away. Immensely powerful, a countryman, a Norman, the Angel swept on as though the weight he carried were no more than a straw. Rémy, following, stumbled and strained, while behind him he heard the imprecations and the sobbing breath of Julitte. At the top of the hill Rémy thankfully paused. He was sweating 'as though in a Turkish bath' and 'my sweat had triggered off the action of the cotton-wool impregnated with mustard etcetera. It burned me fiercely, back and front, giving the impression that I moved in a ball of flame.' He longed to tear off his carapace, but could not reach it under its layers of wool and leather protection. In any event, it was extremely cold on the snowy plateau, and he could imagine he heard Edith saying, 'Don't you dare take anything off.'

Rémy's watch showed 11 p.m. German time, the equivalent of nine o'clock Greenwich Mean Time, and that was supposed to be the latest limit for the operation. But they waited, thinking of the effort it was going to take to carry all those sacks and cases back to the café. Julitte was beginning to play up again. 'Nous en sommes pour nos frais,' he said. 'What fools to imagine that at last . . .' But the Guardian Angel had held up a hand the size of a sucking pig. He had heard an aeroplane, far away, a single small aeroplane. Rémy suddenly became aware that his terrible cold had completely gone, vanished, blown away, adieu! Bob had switched on the three electric torches on sticks, and was sending the code letter in morse with a fourth torch.

Losing height rapidly, the Lysander swept round in a circle, switched on his landing light, touched his wheels on the base of Bob's triangle, rolled on to the apex, turned and taxied back to swing into the wind at the base. From the dark fuselage a figure descended. Rémy had known that an agent called Anatole was coming, and that the Guardian Angel was to give him a bed for the first night. Anatole proved, though, to be a woman, a young woman with fair hair. The Lysander's engine still turned. The sacks and other baggage were flung in. Julitte went up the ladder. The Guardian Angel seized Rémy in his muscular arms and brushed his cheek with a sharp bristle of moustache. As Rémy climbed the metal ladder, a bottle of the Angel's Calva in one hand, he heard Julitte shouting from the interior, 'Look out, there's no room in here.' But Rémy pushed himself in and somehow sat down despite Julitte's protests. The Lysander was already taking off. They pulled the perspex canopy shut above them. The aircraft climbed masterfully over the wood, over fields with the outlines of hedges.

Julitte nudged him. 'Look.'

They were above what appeared to be an aerodrome near the edge of the sea. A light ran rapidly across the surface below them, from one edge to the other.

'A Boche fighter!'

Seeking to warn their pilot, Julitte hunted in vain among the piles of their belongings for the inter-com. He failed to find it. They were well out over the Channel. The sky seemed empty save for themselves. Rémy in gratitude looked at the stars. He began to pray. . . . [2]

'Je vous salue, Marie, pleine de grâces . . .'

He was half asleep, thinking of his other journey to England, from Brittany with Claude in 1940, when a brilliant red light passed the length of the cockpit. The pilot had fired a flare as a recognition signal, and below them were the chalk cliffs of Dover. The Lysander tilted far over on one wing, straightened, and landed at Tangmere. What Rémy did not know while they flew over the Channel was that down to the westward the Whitleys of 51 Squadron were lumbering in the opposite direction, carrying 'C' Company with its section of Royal Engineers, Flight-Sergeant Cox and 'Private Newman' on the Bruneval Raid.

## NOTES TO CHAPTER 18

1. Agents had to be trained as to the dimensions and surfaces required from clandestine landing grounds, setting out of lights, flashing of the code letter, security, and so on. See Foot, *SOE in France*, and Cowburn, *No Cloak, No Dagger*.
2. Rémy, *Bruneval*, p. 165.

# 19

# 'I Feel Like a Bloody Murderer'

*February, 1942*

When he had seen them off to the aerodrome in their trucks Frost sat down to dine with the glider pilot officers. He felt it hard to say nothing of the speculation and fears that galloped through his mind. He looked round him at the placid faces. Would he ever see them again? Lucky devils! Going soon to bed in their warm hut.

After dinner he went to his room and dressed in full parachutists's rig. It felt even more cumbrous than usual. He returned along the passage, and could not resist the impulse to open the door of the Mess and poke his head in. The pilots were dozing peacefully near the fire. His soldier servant waited on the doorstep beside the Humber staff car. They both got into the back seat with some difficulty, bulky as Victorian ladies. The drive to Thruxton seemed pleasantly attenuated.

'The Company was dispersed in huts round the perimeter. John Ross and I and Sergeant-Major Strachan and Newman visited each little party in turn. Some were fitting their parachutes, some having tea. There was a lot of talk, and one group was singing. It was a glorious night, an utter change in the weather; the kind of thing that only happens in England. Sometimes an aircraft engine would cough into life, and always in the background there was the rattle of a truck coursing round the edge of the aerodrome on some urgent mission. That truck came round again, and it was looking for me. I was wanted on the telephone. Were they going to call it off again?[1]

'But it was Group-Captain Sir Nigel Norman. "Just want to say good luck, Frost," he said. "Latest information is that there's snow on the other side, and I'm afraid the flak is lively."

'So the RAF had been snooping round there again, annoying what we wanted to be a sleeping hive of Germans. They had promised diversionary raids by Fighter Command on neighbouring parts of the French coast; it was to be hoped that the weather report came from one of them. As for the snow . . . in a way it was a relief to me. We had been issued with snow smocks but they were back at Tilshead. No, my feeling was that our main difficulty on such an operation was the "fog of war"—confusion. I thought the snow would make things clearer.'

Flight-Sergeant Cox found the night scene dramatic. 'We were put in blacked-out Nissen huts.[2] Inside it was warm and the light was yellow. Parachutes were laid out in rows on the swept floor and we each picked one hoping that the dear girl who'd packed it had had her mind on the job. These were dark 'chutes, camouflaged in greens and blacks. Until then I'd only used white or yellow ones. They pressed bully sandwiches on us, real slabs, and mugs of tea or cocoa laced with rum. We checked each other's straps, and wandered about wide-legged, like Michelin men. It seemed brighter outside than in the hut, and bitterly cold. We were formed up in our tens, or jumping sticks as we had done so often in training. One saw then what it had all been about. It was reassuring to know exactly where to go and who with. Piper Ewan was playing, and that has an effect on Scotsmen. It brings them to the boil—and they're excitable enough already. I'm not sure that the pipes are healthy. The piper was coming on our jaunt, but leaving his pipes behind.'

Ewan played their regimental marches as they moved round the tarmac to the twelve Whitley bombers. Frost was exhilarated by the pipes and impressed by the airmen. 'They were a different breed, at ease, and dressed for what they normally did. By comparison we seemed a lot of clowns. I had a waterbottle of strong tea laced with rum, and I handed it round the blokes in my stick while we waited to emplane. Charles Pickard came along to see me, puffing (unlawfully I suspect) at his pipe, a reassuring and a wholly proven leader, and very young at that. I wasn't flying with him. He was leading the flight, with Junior Charteris's lot. When he drew me aside from the others I expected some joke or platitude. But he certainly made no attempt to reassure me personally.

'"I feel like a bloody murderer," he said.'

'I was jumper number six in aircraft number six,' Cox remembers. 'We put on our silk gloves and crawled into sleeping bags for warmth. The Whitley's ribbed aluminium floor was fiendishly uncomfortable. Ahead we could hear a kite revving prior to take-off and then it was away. Others followed until it was our turn. The whole machine throbbed and bumped and dragged itself off the ground as though it had great big heavy sloppy feet. Nobody slept in that dim-lit metal cigar. We had some singing. We sang *Lulu*, *Come Sit by my Side if You Love Me*, and *Annie Laurie*. Then, by popular request, I gave them two solos, *The Rose of Tralee*, and *Because*. Somehow the engines made it easy to sing. They came in thrums, noisier one instant than the next.'

Corporal Stewart and two others had played cards all the way over. Stewart was winning as usual. He pulled out his wallet to put in yet another bank note and said generously that if he copped it on the raid the bloke next to him must take the cash and make good use of it.

* * *

*Prins Albert* was steaming fast back to Portsmouth. Her landing craft, escorted by the motor gunboats, were closing the French coast under their own power. There was as yet little wind (though more was forecast quite soon—too soon perhaps), and visibility was good, with a bright moon, little cloud and some haze.

* * *

When the cover was taken from the hole in the floor a piercing blast entered. Those who cared to look down saw the sea, gently moving in the moonlight. Suddenly they were over snow-covered land, bouncing and weaving in anti-aircraft fire. Frost dangled his legs in the hole. His bladder felt ready to burst from all the tea drunk at Thruxton. He was so uncomfortable that he could not wait to get down. His companions were in a similar plight. He swore that he'd never get caught like that on any subsequent operation. Why had they not been told to keep off the liquids? Or indeed why hadn't he thought of it?

'Action Stations . . . GO!' One after the other they shot out. As soon as Frost's parachute opened he saw below him all the Dropping Zone landmarks, so well known from hours spent with the model. They were dead on target. He landed very softly in snow a foot deep. No wind. No German reception committee. The aircraft were swimming away over France and there was flak from the main Radar station, probably firing at the last of the Whitleys coming in.

Cox lay in the slipstream looking at the dark belly of the aircraft.

How fat it was, and how fiery the points of its exhausts! Suspended, now, from his parachute, he felt to check that he had his fighting knife and his Colt ·45 automatic. Then he was rolling in the snow. 'The first thing that struck me was the hush. I suppose it was reaction to the horrible din inside the aircraft. Then I heard a rustling, and saw something outlined on the snow and a light on it.' He disconnected the container's light and that of the hand-trolley, which had come down near by.

Having got rid of the Thruxton tea, 'not good drill but a small initial gesture of defiance' (Frost), they gathered at the line of trees. As they did so Timothy's RODNEY party were coming down in a sizeable cloud of parachutes. With such visibility, Frost knew, there was little chance of surprise, and anyway, presumably HENRY would have tracked them across the cliffs. The one good thing was that, even if a lot of Germans now knew that he and his Company were there, they would have no idea where and how they were going to be hit. Everything seemed to be going so well that John Ross's news was a shock.

Ross and his heavy section, the organisational rearguard of the beach assault party had landed safely with their gear. So had another (Charteris) section of ten men, but it had a special task, and it left immediately to do it. The task was to take and hold the German pillbox (REDOUBT) on the north side of, and above, the beach. Should the pillbox be manned they would need surprise, luck, and dash to take it. Having taken it, they were expected to act as the pivot for the whole withdrawal and also for the assault on the beach defences. (In the event they found the pillbox unmanned, but quite rightly remained in it according to orders.). *Euan Charteris was missing with his other two assault sections, twenty men in all.*

Not for the first or the last time, Frost inwardly cursed the inflexible plan that had been thrust on him. This was exactly what he had feared would happen. Two aircraft had failed to deliver their parachutists. Perhaps they had been shot down, perhaps they had dropped them somewhere else. So?

He asked John Ross to wait for a few minutes to see if Charteris and his men turned up, and then to get on down to the beach defences and to do his damnedest with the heavy section. The main problems, HENRY and LONE HOUSE, must be tackled first as planned, and as soon as he could he would get Peter Naumoff down to help Ross at the beach. Meanwhile Naumoff and also Timothy's lot were needed to hold off any attack on the raiders at HENRY, either from RECTANGLE or from Bruneval village.

Within ten minutes of landing No. 2 Party had formed up into its four components. Naumoff led his people off in the RECTANGLE (Le Presbytère) direction; Frost, with Newman at his heels, led his towards the plainly visible LONE HOUSE; and Peter Young and his assault section made for the *Würzburg*, followed at a slight distance by the engineers and Cox, wheeling the trolley.

To Frost's astonishment, the front door of the house stood wide open. The hall was dirty, empty, quite unfurnished. He could just see that Young's party were round HENRY. Young and Sergeant Mackenzie and three others would have hand grenades ready, the pins out. The plan was to fling the grenades as soon as the whistle went, then charge in with their Stens. Frost blew his whistle four times and darted inside. The ground floor was empty, but shooting came from above. They ran upstairs and found only one German there, firing down at the tremendous shindy going on round HENRY. They killed him, and searched the rest of the house. It was empty.

Meanwhile Young and his men had overrun the *Würzburg* position and those Germans who could took to their heels. One of them scuttled towards the cliff edge, the moon-dazzle on the sea silhouetting him. There had been too much shooting, Young felt, and no prisoner had yet been taken. The German was chased and he fell over the edge of the cliff but managed to cling on and find footing. As he climbed back he was caught and taken to the Radar bowl. He was unarmed.

Dennis Vernon had left his own men and Cox kneeling in the snow. He went forward to reconnoitre and after a few seconds they heard him call, 'Come on, the REs'.

Cox saw that the barbed wire round the pit was low and not much of a barrier. He thought it had probably been kept low to avoid electrical interference with the set. The firing from the big house had now stopped, mercifully, but more firing was coming from another direction, RECTANGLE. Major Frost soon appeared at the Radar pit and Newman was questioning the badly shaken German prisoner. He confirmed that he belonged to the Luftwaffe Communications Regiment, and that there were about a hundred of his fellows quartered across the fields at Le Presbytère (RECTANGLE). They were fully armed for defence of the main *Freya* position and also of the *Würzburg*. Yes, he said, in reply to Frost's question, yes they had mortars, but were not in the habit of firing them much, they being signallers. Naumoff and his section were responding steadily to the fire from RECTANGLE, which was mainly directed at them and at the house, rather than at HENRY. Frost also heard firing from Timothy's lot, farther inland. His own group and Young's took up closer defensive positions.

Cox tore aside the thick black rubber curtain that shielded the entry to the Radar set. 'Hey, Peter!' he called to Newman. 'This thing's still hot. Ask that Jerry if he was tracking our aircraft as we came in.' The prisoner agreed that this was so. The *Freyas* in the main part of the station had picked up the British aircraft far offshore. This set had picked them up at thirty kilometres. They had expected to be bombed, and had been getting worked up as the hostile machines came in low and virtually straight at them. The signaller pointed out that the site was 'extremely exposed'; on learning that the hostile aircraft were making directly for them, they had switched off in good time, and had taken cover.

Vernon began to take flashlight pictures of the *Würzburg*, while Cox made notes and sketches, using his hooded torch. The flashlights at once drew German fire, and Frost ordered him to stop the photography.

'Like a searchlight on a rotatable platform mounted on a flat four-wheeled truck,' wrote Cox. 'Truck has had its wheels raised and is well sandbagged up to platform level. Paraboloid is ten-foot diameter and hinged so that radio beam can be swung freely up or down or sideways. Small cabin to one side shelters set's display gear and operator's seat. At rear of paraboloid is container three feet wide, two feet deep, five feet high. This appears to hold all the works with the exception of display. Design very clean, and straightforward . . ., We found the set switched off, but warm. The top of the compartment taken up by the transmitter and what looks like first stage of receiver. Large power pack with finned metal rectifiers occupies bottom. Between the TX and the power unit is the pulse gear and the receiver IF. Everything solid and in good order. Telefunken labels everywhere which one sapper was removing with hammer and cold chisel. Just enough light to work by, with moon reflected off snow.'

Finding that there was no quick way of removing the aerial element in the middle of the bowl, Vernon ordered one of his men to saw it off. He agreed with Cox that the important material was in the consol, rather than in the display. They removed the pulse unit and the IF amplifier in a civilised way, using good tools on a well-maintained machine. They then tried to get out the transmitter. Cox had an immensely long screwdriver, but it would not reach the fixing screws. He and Vernon conferred, then the two of them grasped the handles and body of the transmitter while a third man put his weight on a crowbar. It was in a light alloy frame. It came away with a tearing sound, bringing its frame with it.[3]

'A stroke of luck,' Cox says. 'When the equipment was examined later it was found that the frame which we in that somewhat hasty moment regarded as no more than an encumbrance, something we had not the time to detach from the transmitter, contained the aerial switching unit that allowed both the TX and RX to use the one aerial, a vital part of the design of an Radar set.'

Crowbars were used to rip out the last of the wanted components, and the engineers frequently had to use their torches. Enemy fire from RECTANGLE was getting heavier and more accurate. One of Frost's party, Private McIntyre, was killed near the door of LONE HOUSE. At last the REs were loading the trolley, Frost was glad to see. He had confidence in Vernon, and he felt that now, if they could only get away with the swag they had won the day. At the same time the battle situation was one of confusion. They had certanly stirred something up! There was firing from nearly everywhere. Heavy firing from down by the beach, with the odd white flare (which must be German) going up. That firing was mainly the deep stutter of a machine-gun, and it was being answered by one Bren. The firing from RECTANGLE had increased and had spread. Obviously the people there had deployed, and soon they would probably advance on HENRY and LONE HOUSE. Then there was an extraordinary amount of firing that seemed to come from the village itself. Had some of Timothy's men gone berserk and fought their way into the centre of Bruneval? Lastly there was a lot of noise on Timothy's front, but a runner had just come in from there. John Timothy said reassuringly that everything was under control except his No. 38 wireless set. That was the main trouble. Frost, except for whistle signals and runners, was without communications, and under the plan laid down he did not have a full complement in his Company Headquarters. As for the 38 sets, they seemed to be quite useless, and would not keep on net at all. The trolley was loaded. He sent a runner to call in Peter Naumoff.

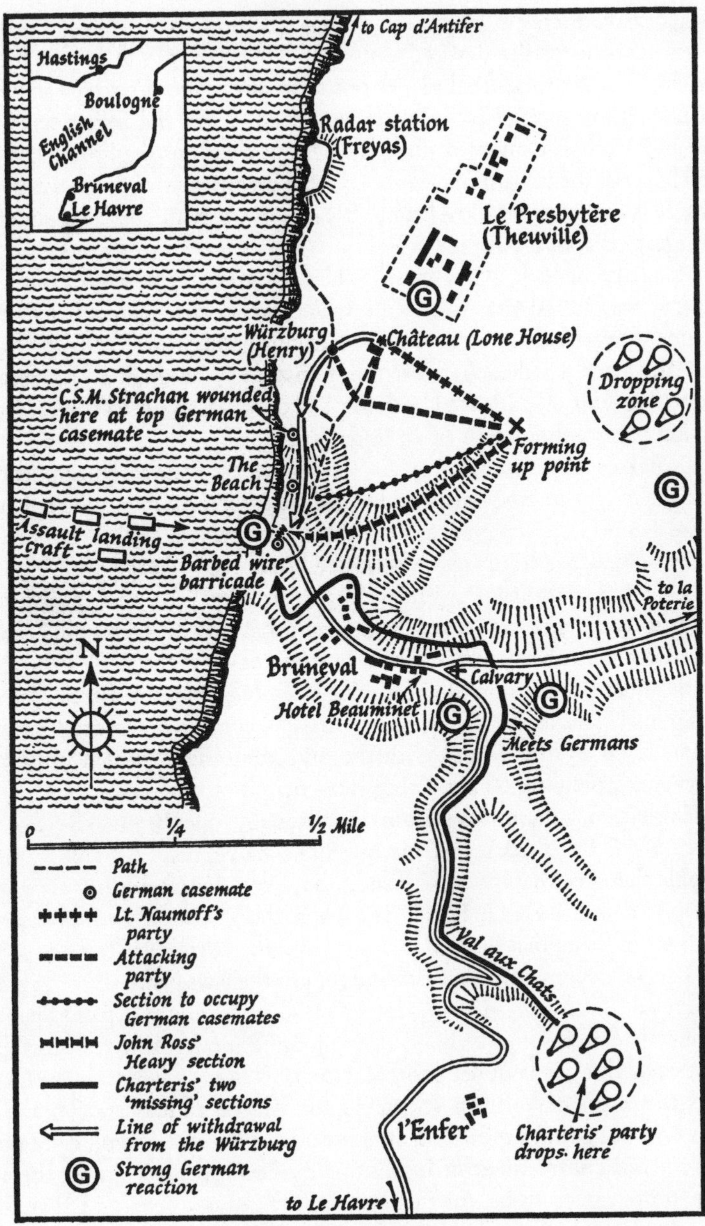


Map illustrating the Bruneval Raid

He was to fall back through them and lead the way down to the beach. 'Remember the password, BITING,' he warned the runner. Young Naumoff had been a supernumerary officer on training. Frost had allowed him to come on the raid because he had not the heart to disappoint him. He seemed to keep very steady. But he had been under fire for some time, and a man who came up from behind might be in danger.

Captain Ross waited until all of Frost's group had left the forming-up place. Then he led his section across ground remembered from the model, and down to the road and the entry to the beach.[4] The plan drawn up in England had envisaged a three-point initial attack. The assault section that had already gone out was to occupy the important hinge pill-box. Ross assumed that they *had* occupied it, as he had heard no firing from there. Ross and his heavy section were to be in reserve in the centre, nearest to the road entry to the beach, while Euan Charteris and his two assault sections were to sweep in from the south-east. As Ross slid from the trees down the hill towards the entry a white flare rose from GUARDROOM. He and the men behind him were at once pinned by machine-gun fire. Lying on the snow, Ross made out that the fire was coming from the inland side of the Villa Stella Maris (GUARDROOM) where, slightly above the road, the Germans had made weapon pits with trench communications. Several rifles, probably six or seven, were firing as well as the machine-gun. Ross's section replied with rifles and their one Bren. The section consisted of himself and his batman, one reserve sergeant, two signallers with a No. 38 set and a No. 18 set for contact with the Navy, two sappers with mine-detectors and a Rebecca radio beacon, two Bren-gunners and one runner. The Germans had an excellent field of fire, and they were hidden by the villa from the British section in the pillbox to the north. It had not been foreseen that the Germans might dig themselves weapon pits on the landward side. The heavy section was pinned down. The sergeant, dragging himself flat on the snow, managed to get to the thick barbed wire across the road, and tried to cut a way through. The Bren-gunners and riflemen kept firing. Ross would have given a lot to have in his section a mortar, even a two-inch one.

Mortars were in Frost's mind too, as he thought of his men concentrating near and on the beach. So far he had heard no bombs, only rifle and automatic fire from the enemy.

In the centre of the withdrawal across the plateau and down the hill was the trolley. The two-wheeled affair took some controlling, being heavy-laden. As Naumoff and his section withdrew through the main party Frost warned him that he would probably have to contact Ross and fight his way to the beach. Naumoff got down unobserved but when the trolley party and its guards came to the German casemate high up on the north side of the gully the machine-gun below caught them in full view against the snow and gave them a long burst. Company Sergeant-Major Strachan fell with three wounds in his stomach. Just then they heard John Ross clearly.

'Don't come down,' he shouted. 'The beach is not taken yet.'

Frost got Strachan behind the concrete jut of the casemate, put field dressings on his ugly wounds, and gave him a morphia injection. A runner came from John Timothy's party to say that the Germans were advancing from RECTANGLE and had already occupied LONE HOUSE. Frost accordingly told the trolley party to hold on by the casemate until the machine-gun below had been silenced, and taking every man he could, including sappers, he hurried back to the top to counter-attack in conjunction with Timothy's group. Whatever was happening down below at the beach, it was vital to keep the Germans above at a good distance. 'Fortunately,' he says, 'the threat did not amount to much. The enemy was confused and did not know what he was up against. They hesitated and withdrew.' Frost left Timothy to defend the shoulder of the hill while he hurried back to the casemate and to take control of the fight for the beach. But he found that Vernon, Cox, and two sappers were on the move down, skidding and sliding on the frozen path. Sergeant-Major Strachan, shouting incomprehensible orders, was being half-carried, half dragged after the Radar booty. There had been a lot of shooting down below, but now there seemed to be silence. Frost dashed on down, leaving the trolley above.

When Ross saw the trolley party appearing on the skyline, and full in the field of fire of the Germans by Stella Maris, and called out to warn them, he realised that the distraction might be turned to account. His men found a knife-rest in the perimeter wire and pulled it aside to make an opening. At that moment Naumoff and his section reached the bottom of the slope, and as they and the heavy section were about to rush through the wire and assault the German position they heard shouts and firing from the south-east. Charteris! The Scots voices raised in anger and triumph at reaching their objective came clear on the frosty air. The attack from the south-east had turned the enemy's position. The German machine-gun was abandoned and the defenders slipped out of their trenches and off to the southward, to the obscurity of the rough ground at the edge of the sea.

* * *

Pickard and the aircraft that immediately followed his had been caught in the worst of the flak. They had mistaken their landmarks and had dropped their two sticks of parachutists well south of Bruneval, indeed almost halfway between Bruneval and St Jouin. The twenty men landed with their containers in the Val aux Chats, near the small hamlet of l'Enfer. Euan Charteris was an outstanding young officer, remarkably intelligent and of the greatest promise.[5] I don't mind saying that it was a nasty moment,' Charteris said.[6] For when he picked himself up from the snow he saw at once that their pilots had made a mistake. Where were they? Fortunately they were able to watch the subsequent Whitleys flying in, well to the northward. They soon distributed the contents of their containers and Charteris's two scouts reported the narrow road leading to Bruneval, which, Charteris realised, lay between them and their objective. Putting himself at the head of his men he led them at a jog trot up the side of the road. As they neared Bruneval they saw other soldiers, but in the half-light they proceeded for some time unchallenged, and when a German joined their line thinking they were his own people he was killed silently. Then the challenge came and they had to shoot their way round the village. This was the firing that Frost had heard while HENRY was being dismembered. But if the firing was confusing to Frost it was still more so for the German garrison, who had little idea where the 'commandos' were or what their objective could possibly be. Now fighting in the half-light, now hurrying on round the village in its sharp valley, now separated, now together again, Charteris's sections shed a man or two here and there. The main party's firing was ahead of them now, and up over the hill on their right. Charteris led them at the double over the road from the sea to Bruneval and turned right-handed for the German defences and the beach. They paused to get back some of their breath, then, mad with relief at being at last where Orders had said that they should be, they charged forward with a wild yell. Their attack came, by a fluke, in conjunction with the attack from the other side of the valley. The three groups, Charteris's, Ross's, and Naumoff's were through to the beach.

At the door of Stella Maris Sergeant Jimmy Sharp caught a German telephonist coming out, and explained to him that he would be making a trip, 'nach England'.

But would he? Ross's two signallers seemed quite unable to make contact with the Navy on their No. 18 set, though they were still trying. The signallers who had been in Charteris's party and who also carried a No. 18 were both missing. As no contact could be made with the No. 18, Ross told his other signaller to keep trying with the No. 38 set. Meanwhile the sappers had set up the little portable radio beacon, the Rebecca, and said it was working properly. It was a gadget so new and so secret that it contained its own built-in demolition charge. Its companion set, known as a Eureka, was in one of the landing craft. Whether the gadget was working or not, the parachutists doubted. Meanwhile Ross's two sappers checked the beach with their mine detectors getting negative results. And Ross himself at his report centre totalled the losses. There were two confirmed dead, Privates McIntyre and Scott, six missing, and six wounded who had all been gathered on the pebbles. Among them was Corporal Stewart, who had fallen, hit in the head, during the assault on the beach. He had called to his nearest friend, 'I've had it, Jock. Take my wallet.'

Lance-Corporal Freeman took it and examined Stewart's head. 'Och it's only a wee bit of a gash,' he said.

'Then give us back that wallet.' Stewart managed to get to his feet.

After consultation with Frost, John Ross fired one green Very light from the north end of the beach and then another from the south. Frost 'with a sinking heart' called his platoon commanders, and began to organise the defences of an indefensible position. There had been reports from Timothy that headlights were approaching from the east and south-east. Before long the Germans were bound to appear in strength.

Bruneval was, for the light naval units, an easier target than the dark rocky beaches of Loch Fyne or the swelling cliffs round Lulworth. The trouble, as the Navy was only too well aware, was that the deadline for the raid had been extended and the landing craft would now approach a dangerous beach on a falling tide. Another difficulty was the weather itself which had become, frankly, unsuitable. By midnight the barometer was dropping sharply and the breeze was already fresh from the south-west, and increasing. While they waited in a state of increasing tension the flotilla of small craft saw enemy ships a mile or so to seaward, between them and England, two German destroyers and two E- or R-Boats. The Germans steamed very fast from north to south and apparently saw nothing. Soon after they had gone the white flares fired by the defenders of Stella Maris were seen. By two thirty-five the landing craft had closed to within three hundred yards of the beach when 'a blue lamp signal was seen followed by two green Very lights'. Two LCAs were ordered to close the beach, but as they started inshore Ross's signaller made contact with his No. 38 set and asked, without authority but understandably enough, that all the boats should go in.

Frost, his back to the sea and his hopes at zero, heard a cry behind him, 'Sir! the boats are coming in . . . God bless the ruddy Navy!' Then the hinterland of the beach where he was standing and where most of his men were deployed was swept by a devastating fire from the Brens in the landing craft.

NOTES TO CHAPTER 19

1. Frost's written account.
2. Cox's written account.
3. Accounts of Vernon and Cox.
4. Ross's account.
5. He was killed fighting with the Second Battalion when, commanded by John Frost, it survived, but only just, one of the most desperate actions of the whole Tunisian campaign.
6. Saunders, *The Red Beret*, p. 67.

# 20
# Return

*February, 1942*

After he had managed to prevent the slaughter of his men by their friends, Frost turned his attention to the difficulties of embarkation. He felt that they had been increased by the simultaneous arrival of all six boats. However, the abortive exercises on training had taught the parachutists more than they realised. Even with the swell and the falling tide, there were enough of them to hold the boats stern-on to the sea. All got away safely who were on the beach. The total left behind was still eight, two dead and six missing. The first landing craft away took the wounded and Cox and the Royal Engineers with their booty. They were transferred to MGB 312, which then made off at more than twenty knots for Portsmouth.

By three-thirty the remaining gunboats were under way, towing the landing craft. The parachutists were tucked away below decks. In the mounting sea with their tows they could make no more than seven knots, and at dawn the flotilla was only fifteen miles from the French coast. The wind had increased on the Beaufort Scale to Force 5. An hour after dawn they were protected by Spitfire cover overhead, and earlier by four Free French chasseurs, *Bayonne*, *Calais*, *Larmor*, and *Le Lavandou*, and by two destroyers, HMS *Blencathra* and HMS *Fernie*. On boarding his gunboat Frost was distressed to learn that the two missing signallers had just made contact, using their No. 18 set. They had reached the beach, and now would have to attempt to escape to Spain or Switzerland.[1]

An excellent sailor (even in a fast motorboat lurching slowly over a confused sea), Frost was dealing with an early lunch ('the sailors certainly upheld the traditions of Royal Navy hospitality; they made much more fuss over us than we deserved'), when a message came for him from NOAH, far ahead in MGB 312: 'Samples complete and perfect.'

Priest had lost no time in examining the bits and pieces and in questioning Vernon and Flight-Sergeant Cox, who was very seasick. 'Once we got home official opinion had it that the German set was behind our own gear,' Cox says. 'For my part, and I told Mr Priest so, I thought it a beautiful job, and I've been in wireless all my life. I was particularly struck by the ingenious way it was boxed off in units for easy fault-finding and quick replacement. "The Jerries must have had RDF as long as us," I said to Mr Priest, "or longer." "Well," he answered, "as soon as we get your evidence home we should know how long they've been making this *Würzburg*, and I'm willing to bet the total will be in years, and nearer ten than one. But keep it under your hat, my friend." When he had done with me I managed to hold down a cup of good strong tea and I went to sleep in the Captain's bunk. Yes! The boat no longer vibrated when I woke, and I found we were alongside *Prins Albert* in good old Pompey harbour. Next morning I reported as per instructions at the Air Ministry. "Take two weeks leave immediately, Sergeant Cox," Air Commodore Tait said to me, so I asked his secretary WAAF to send a telegram to Wisbech saying, HOME TONIGHT KILL FATTED CALF. It was near midnight when I got home, but sitting round a big fire there they all were, four generations. "Hullo family," I said. "I've been in France, that's where I've been, and it's in the London newspapers tonight. How about that then?"'

John Frost went on the bridge as they approached Portsmouth. Those were moments that this young soldier, with a military career in front of him that held more than a fair share of danger and glory, would not forget. The four French chasseurs now swept by and saluted the raiders, followed by the two British destroyers; and the Spitfire escort, its task acccomplished without enemy challenge, flew low over the gunboats before making off inland. At six that evening the survivors of 'C' Company, its section of Royal Engineers, its Flight-Sergeant, and its German interpreter boarded *Prins Albert*. The ship was crowded with staff officers and journalists. Pickard and the Whitley crews were there.

* * *

When the Lysander carrying Rémy and Julitte had landed at Tangmere 'J', a mysterious British officer who had briefed Rémy in Portugal at the beginning of his career as an agent, was there to meet them. He led them to a room where a woman in Army (ATS) uniform sa touching up a coke fire with a steel poker. Rémy and Julitte were given whisky, more whisky, then bacon and eggs, toast, marmalade, and coffee. A black car driven by a woman whisked the three men to London, where J installed Rémy in the Waldorf Hotel under terms of the strictest security. False papers had been made out. The reluctant Rémy was a 'French Canadian named Georges Roulier'. He was ordered to take all his meals in his suite, rather than descending to the restaurant. When J at last left him, Rémy plunged into a hot bath, at that time a rarity in France.

Early next morning he was taken to Passy's headquarters in St James's Street. His reception was enraptured because all those papers from the Lysander had already been appraised. Rémy permitted them to give him (strictly against J's security instructions) luncheon at *L'Ecu de France* in Jermyn Street. J was waiting for him at the Waldorf, and handed him as a surprise an early edition of the *Evening Standard* whose headlines and whole front page were concerned with the successful raid at Bruneval.

Utterly delighted, and thinking as usual of others rather than of himself, Rémy sat down at the desk in his private sitting room and wrote a message for Pol that J undertook to have transmitted immediately.

> TO PACO FOR POL CONGRATULATIONS SUCCESS BRUNEVAL WHICH HAS RESULTED DESTRUCTION IMPORTANT GERMAN INSTALLATION WHILE TAKING AND KILLING NUMEROUS BOCHES

'Paco' (François Faure) was one of the original and most important members of Rémy's Confrérie Notre Dame. Rémy was bitterly to regret that moment of exhilaration in the Waldorf.

Two days before the Bruneval Raid, in the Battle of the Java Sea, Anglo-American forces, so over-optimistically deployed following the meeting in Washington after Pearl Harbour, had been almost eliminated, and the Japanese occupation of Java made sombre reading. Accordingly the news of Bruneval, the first successful armed landing in German-occupied Europe, was a tonic for the Allies and a blow to German pride and confidence. Lord Haw-Haw, the English traitor propagandist, referred scathingly to Frost and 'C' Company as a handful of redskins. And, possibly for security reasons, the decorations awarded to the parachutists were sparse, even by British Army standards. Frost, a professional soldier, and Charteris, whose ambitions were political rather than military, each received the Military Cross, while Cox, Sergeant Grieve, and Sergeant MacKenzie received the equally distinguished Military Medal; Young was Mentioned in Despatches and Company Sergeant-Major Strachan, who lived to fight again, but only as a frail shadow of his former self, was awarded the Croix de Guerre.

From the purely military point of view, Combined Operations under its new leader, Mountbatten, had tackled something dramatic and difficult, and had carried it through well. Churchill was impressed, and his appetite for raids, always voracious, was whetted. Only one of the Bruneval parachutists took part in the next raid, the extremely gallant and effective one on St Nazaire. 'Private Newman' had volunteered his services once more as German interpreter. He was taken prisoner, and spent the rest of the war in German hands. But so good was his cover story that his real nationality remained a secret, and he survived the war to become a prosperous business man in England. The third important raid on France before D-Day was that on Dieppe, a ferocious affair of much value in view of the great landings that lay over the horizon.

* * *

German reaction at Bruneval during the raid was described in the following report,[2] which revealed that the British could scarcely have chosen a worse night to attack . . .

> At 0055/28.2.42 the German *Freya* station reported aircraft NNE, range 29 km.
>
> The parachutists were sighted by the Army and the Luftwaffe (ground and communications troops) at 0115. The landing was made SE of the farm and was carried out in complete silence.
>
> All Army and Luftwaffe posts in the area were at once alerted. Scouts sent from the *Freya* position (near Cap d'Antifer) and the Luftwaffe

Communications Station (at Le Presbytère) returned with information that the enemy was on the move south of the Farm (Le Presbytère) in the direction of the Château. The parachutist commandos had split into several groups and were converging on the *Würzburg* position and on the Château.

In La Poterie the reserve platoon of the First Company 685 Infantry Regiment had just finished an exercise shortly after 0100 when the parachutists were sighted. The officer commanding at once made contact with the Bruneval Guard; the Sergeant there had already alerted his men.

The platoon reserve in Bruneval was ordered to occupy Hill 102 to the SE of Bruneval. The officer commanding La Poterie platoon then led his men in a westerly direction towards the Château.

On reaching the Farm buildings NE of the Château the German troops came under fire from the commando machine-guns, and from the W end of the buildings they engaged the British, who were already in possession of the (*Würzburg*) Luftwaffe station near the farm. Here one of the commandos fell.

This German platoon encountered fire from the left flank, but the commandos were nevertheless prevented from proceeding with their attack on the *Freya* position. The remainder of the Luftwaffe Communications Station unit quartered in the farm buildings took part in this action.

In accordance with orders, the platoon from Bruneval village divided into two groups and advanced on Hill 102. Outside Bruneval they came under fire from the commandos who had landed N of L'Enfer.

Although this platoon was unable to prevent the commandos from infiltrating between Bruneval and Hill 102, it was because of this platoon's action that individual commandos did not reach the boats in time, and were later taken prisoner. One wounded commando was also captured. It was only because the British objective was not known that this Bruneval platoon did not take part in the action at the Château.

The Bruneval Guard, one Sergeant and nine men, had meanwhile taken up prepared positions guarding the coast. These (main) defensive positions were so built that they were effective only against attacks through the ravine from seaward. The commandos, approaching from N and NE were able to get close to these strongpoints under cover of the woods. Thus the German guard positions were attacked from the high ground by heavy fire from three or four commando machine-guns. After one German soldier had been killed and another wounded the Sergeant was obliged to take up new positions. It was not until after one to one-and-a-half hours fighting that the commandos were able to get through the strong-point and the ravine to the beach. With them the commandos took a wounded German soldier and also the soldier who had been on telephone watch at the post. Here another commando fell, and one was wounded. The latter was assisted to the boats, which had come close inshore on the exchange of signal flares.

The commandos embarked just as strong German reinforcements reached Bruneval.

The platoon from La Poterie fought their way to the Luftwaffe Communications Station (*Würzburg*) as the commandos withdrew. It was learned that the Luftwaffe personnel there had put up a stiff resistance and, only after some of them had exhausted their ammunition were the commandos able to break through to the *Würzburg*.

One of the crew had been killed by a British grenade as he tried to set off an explosive charge to destroy the *Würzburg*. The commandos then dismantled parts of the set and also took photographs. On conclusion of this task they obviously intended to attack the *Freya* station. The skilful intervention of the La Poterie platoon, however, prevented this.

The operation of the British commandos was well planned and was executed with great daring. During the operation the British displayed exemplary discipline when under fire. Although attacked by German soldiers they concentrated entirely on their primary task. For a full thirty minutes one group did not fire a shot, then suddenly at the sound of a whistle they went into action.

German losses: Army, two killed, one seriously wounded, two missing. Luftwaffe, three killed, one wounded, three missing. British losses: two killed, one wounded (reached the boats), four captured.[3]

### NOTES TO CHAPTER 20

1. They nearly succeeded. The following letter from M. Maurice de la Joie was published on February 3, 1946 in the newspaper *Havre Libre*: 'We took in two English parachutists who had failed to get aboard their fast motor boats. After my sister, Mme Delarue, had sheltered them for several days, we had them with us again. On March 9 (1942) we were arrested with them when about to cross the Line of Demarcation at Bléré, Indre-et-Loire. Condemned to death by a German Military Tribunal at Angers, we were taken to Paris, where we were imprisoned, my wife in La Santé, myself in the Cherche-Midi. In January, 1943 we were deported to Breslau in Silesia. Then my wife was sent to Ravensbrück, and I to Buchenwald. We both survived the war, but as invalids.'

   All six of 'C' Company taken prisoner survived the war.
2. German Document TSD/FDS/X.378/51, Cabinet Office.
3. Two of 'C' Company were still at large in France when this report was circulated.

# 21
# Deadly Answer

*1942/43*

One immediate German reaction was predictable, and was a gift to Germany's adversaries. German radar stations on the Channel coast were surrounded with barbed wire, and were given all-round tactical defences. Under the wire aprons the grass, protected from the tongues of the omniverous brown-spotted cattle, grew long and rank, the weeds proliferated. On aerial photographs, in consequence, the *Freya* and *Würzburg* positions showed up like warts. This was to be useful before D-Day when it was vital to the Allied plan that all the enemy Radar stations should be taken-out by rocket-firing or precision-bombing aircraft, or should be completely hoaxed so that the enemy might have no precise warning of the two cross-Channel Armadas.[1]

In terms of the bombing assault on Germany and the Germans' characteristically brave and energetic reactions, the Bruneval Raid had an important effect, and one that had much to do with TRE's saturnalia of spoofing that preceded D-Day.

TRE specialists, using the captured pieces from Bruneval and the photographs and descriptions of Vernon and Cox, built a *Würzburg* at Worth Matravers and began evaluation exercises. R. V. Jones had twice been cautioned for speeding when he first drove down to see the raid's booty. His examination of the metal labels showed, of course, that the apparatus had been manufactured by Telefunken (whose factories near Berlin were then at the outer limit of Allied bombing range). The lowest serial number brought back from France was 40,144, and the highest was 41,093. Jones thought that, according to the German practice with armaments, fifty per cent would go into spares. The earliest inspection date (on the transmitter) was November, 1940, and the most recent (on the aerial) was August, 1941. Jones calculated that five hundred *Würzburg* sets would have been in service by the latter date, and that Telefunken would be turning out a hundred a month.

(This estimate was accurate; but the firm was also now manufacturing the more effective Giant *Würzburg*.)

One scientific deduction from the Bruneval capture was as plain as it was unpleasantly controversial . . . The *Würzburg* could be tuned over a wide range, and was not susceptible to 'orthodox' jamming. If the *Würzburg* was as essential to the German air defences as seemed likely, there was but one known answer to it; an answer that each side had discovered independently and had swept under the top-secret carpet; an answer that had a *Boy's Own Paper* smack to it. For it consisted simply in the aerial dropping of metal-strip reflectors which, by creating hosts of false echoes or reflections in the enemy Radar, would conceal the advancing aircraft.

Each side, believing itself to be the superior in Radar and therefore likely to be the more affected by such a blotting-out process, had been horrified by the idea. Göring had given an order that its German codename, *Düppel*, must never be mentioned, even in meetings attended by such persons as himself, Milch, Martini, Kammhuber, Plendl, Galland, and Engineer-Colonel Schwenke, the 'captured enemy equipment' expert. By 1942 the Germans were at last, and too late, beginning to realise that they were behind in the Radar struggle that must lie ahead. When in 1942 the British began to jam *Freya* and to attack German ground-to-air communications with two airborne TRE-devised methods (Mandrel and Tinsel) the already defensively minded German Air Staff began to wonder in committee when the more important *Würzburg* system, now controlling night fighters, searchlights, and guns, would be attacked.

In London this difficult question, urgently raised by the Bruneval Raid, had been asked by Dr R. V. Jones after a visit to Bawdsey Manor back in 1937. He had noted that the Radar at Bawdsey would detect a piece of wire 'half a wavelength long' (a dipole) hanging from a balloon twenty miles away.

On his return to London, Jones—who was then researching infra-red methods of aircraft detection—expressed this opinion to Lindemann concerning the British Radar defence system: 'All the Germans would have to do would be to sow a field of dipoles over the North Sea, and our screens would be swamped with echoes.'

Lindemann was impressed. 'I'll get Winston to raise it,' he promised.

When Lindemann had explained dipoles to him, Churchill did raise it, in the important Air Defence Research Sub-Committee. He got little satisfaction. Tizard and Watson-Watt admitted to him that scattering dipoles represented a potential way of blinding Radar. But neither they, nor anyone else outside Germany, suspected that the Germans already possessed Radar. Both Tizard and Watson-Watt believed that Britain's survival in the coming war depended on developing the Chain Home and its ancillaries with all possible effort and devotion. And events justified them, showing the Germans to be less aggressive opponents in the radio and Radar war than Jones had anticipated. It was this query, however, as to the vulnerability of Radar that initially shaded Lindemann's, and therefore Churchill's, early opinions as to its reliability. (Hence Churchill's suggestive 'M'yes,' to Wimperis.)

Only a few months before John Frost and his party dropped at Bruneval, TRE had begun in Dorset the first metal-strip-dropping trials. They were ominously effective. Cockburn came into Rowe's office to report, and asked what code-name should be given to the experiments. TRE had had reprimands from Intelligence for being too clever with code-names. This matter was obviously dynamite, and called for something really stupid. Rowe looked round the room, and said, 'How about "Window"?'

Initial experiments with Window were in the hands of a woman, Mrs Joan Curran, at TRE. She found that rectangular strips of ordinary tinfoil made the most satisfactory reflectors. The development was continued under another unusually interesting person, a peace-time Oxford physicist and don. Dr Derek Jackson before the war had ridden his own horse in the Grand National, and during the war had won both the Air Force Cross and the Distinguished Flying Cross as a Radar-operating observer in night fighters. In the course of these protracted researches he nearly lost his life when the Beaufighter from which he was sowing Window was attacked in error by a Spitfire, which shot down the companion Beaufighter, killing its occupants who included Dr Downing, the TRE expert on the new Mark IX fighter Radar.

From the day of the Bruneval Raid until the following June, Jones's office was working its way to familiarity with the German system of air defence. At first they spoke of The Main Belt. Then Jones's number two, Charles Frank, came up with 'The Kammhuber Line', and the name stuck. 'Lines', such as the Siegfried and the Maginot, had been military and journalistic talking points until the beginning of the war of movement, a war whose vital forces were in the air and still, to some extent, on and in the sea; possibly in Jones's office and at TRE The Kammhuber Line was popular because they knew that, although it was scientific, and for a time effective, its days were numbered. Jones himself liked the name; of all his adversaries on the German side, he had taken a fancy to Kammhuber because he was honest, and patently decent. After the war, when Jones interrogated him, he took pleasure in informing him that his defences had been known in England as The Kammhuber Line. 'A warm smile came over his features. I think it almost made up for being in prison,' Jones says.[2]

Understanding of Kammhuber's system showed that Bomber Command should change to concentrated formations. Since there was only one night fighter at a time in each Kammhuber 'box', it would obviously be sound to saturate the box with bombers. Scattered formations were playing Kammhuber's game. Thus began the British 'bomber stream', which in time was going to bring the remarkable German reaction of feeding fighters, like killer fishes, in among the Lancasters, Halifaxes, and Stirlings.

Tacticians in the higher echelon, from Churchill down, argued the pros and cons of using Window. The more they discussed it, the more reasons were advanced for its non-use. Jones, who had been against using $H_2S$ in bombers over Germany (because it meant giving the Germans the secret of the cavity magnetron valve) rather than using it exclusively in the Battle of the Atlantic, was in favour of using Window. Lord Cherwell, who had insisted that $H_2S$ be used at any rate in the Pathfinders of Bomber Command, and who had had his way, was determinedly opposed to using Window. On the other hand Bomber Command's leader, 'Bomber' Harris, had agreed with Cherwell about using $H_2S$, and now agreed with Jones that Window should be used. One of Cherwell's strongest arguments against its use was that the Germans had never thought of it, and that to put the notion before them would be to invite lethal air attacks on Britain. Jones disagreed. He refused to believe that German discoverers and developers of Radar had not independently discovered the effects of scattering dipoles. And in October, 1942, he got a rumour about the German *Düppel* experiments in the report of a British agent in Germany based on a chance conversation with a German WAAF in a train. When Jones told Cherwell of this, the latter exploded. Did Jones imagine that Air Strategy should be affected by what a low-ranking German soldieress had gabbled about during a German train journey? . . . Jones did.

Window was first used seventeen months after the Bruvenal episode, in the most deadly raids of the war, target Hamburg. Thanks to the delay, Bomber Command was able to deal an infinitely heavier series of blows, and the German defences had little time to find an effective answer to the Window techniques. And thanks to Dr (or Wing-Commander) Jackson, there was now an efficient form of Window, a package weighing only two pounds that produced the Radar echo of a heavy bomber. Also thanks in part to him, to TRE, and to the British and American electronics industries, British Radar, both ground and airborne, could itself see through the Window (or *Düppel*) clouds.

In the gloaming on July 24, 1943, seven hundred and ninety-one British heavy bombers clattered and clawed their prehistoric way into the sky from their East Anglian bases and thundered over the North Sea. At the end of the usual briefing a special announcement from Harris had been read to every crew. 'Tonight you are going to use "Window". It consists of packets of metal strips which produce almost the same reactions on RDF as do your aircraft. The German defences will become confused . . . When good concentration is achieved Window can so devastate an RDF system that we ourselves have withheld using it until we could effect improvements in our own defences.' At this stage in the war Bomber Command did not (like the Luftwaffe) maintain strict radio discipline. The German listening service, familiar with the British test traffic, had predicted an unusually heavy raid for that night. German long-range Radar had watched the first bombers taking off, and had reported the forming of the head of the two hundred mile long stream.

Soon the British began to drop Window, one packet each minute from each bomber, down the flare chute in the tail. They were flying at nineteen thousand feet in air twenty degrees (Centigrade) below freezing. The advanced defence stations on the islands of Sylt and Heligoland at once were almost blinded by reflections.

'*They are multiplying! The British are reproducing themselves!*'

At twenty to one the leading bombers crossed the enemy coast. On their $H_2S$ screens Hamburg showed like a veined opal set in diamonds. Hamburg, the most ferociously defended city in the world, with its fifty-four *Würzburg*-controlled flak batteries, its twenty-two *Würzburg*-controlled searchlight batteries, and its twenty *Würzburg*-controlled 'four-posters' served by six night-fighter aerodromes, had been blinded. The blue master-searchlights that usually remained vertical until pouncing, spider-like, on an intruder, were groping about the sky as though demented. Their operators, like the gunners and the fighter pilots, were waiting for directions that could not be given. Undisturbed, the Pathfinders marked it out accurately. Their immense line of followers began to ram in the bombs. As they turned away they continued to sow Window.

Hitler was wakened early and given details of the Hamburg disaster, including the use of Window. He at once ordered full priority for the production of the V2 rocket. By doing so he unwittingly played the Allied game, since the V2 was an insatiable user of electronic components, and these were already drastically short for German Radar development. The rockets' demands were to hamper Dr Plendl, Milch, and Martini in their search for the difficult answer to the clouds of tinfoil.

Four nights later, in the second attack, Bomber Command turned Hamburg into an inferno. What the Germans described as 'fire-storms' raised winds of hurricane force that blew living people about like chaff, carrying them up into the smoke, flinging them into the hearts of shrieking furnaces or into the boiling waters of the lake.

By contrast with such horror, little Charles Cox of Wisbech in Cambridgeshire, standing dismantling a *Würzburg* under fire, the snow around him, the moon above, seems a peaceful and peaceable figure.

## NOTES TO CHAPTER 21

1. The object of the Bruneval Raid was not, of course, to knock out the Luftwaffe Communications Station at Cap d'Antifer, but simply to capture parts of the *Würzburg*. However, the Cap d'Antifer Station acquitted itself extremely well that night and subsequently. Here is an excerpt from the War Diary of the German Admiral Commanding the Channel Coast at the end of May, 1944:

> 'In the last ten days of May there has been a considerable increase in air attacks on Radar stations, especially in the Seine–Somme and Normandy areas. In the whole of May there were 47 such attacks, including 6 on the station at La Pernelle, 6 on Cap d'Antifer, and 4 on Pointe Precée. In spite of the increased intensity of the attacks and a certain amount of damage to Radar stations, a Radar watch covering the entire coastal area of the command is guaranteed. . . .'

The Diary notes further that an air attack on Cap d'Antifer on June 2, four days before D-Day, caused 'slight' damage. 'But the following day the station was again operational.'

According to the War Room records of the Allied Expeditionary Air Forces the Radar station at Cap d'Antifer was attacked by AEAF aircraft on the following dates:

May—7 attacks on 11, 22, 23, and 30 May. In all 196 aircraft dropped 55·4 tons of bombs.

June–10 attacks on 2, 3, 4, 10, 12, 13, and 15 June. 150 aircraft dropped 48·7 tons of bombs.

July—2 attacks, on 3 and 12 July. 6 aircraft dropped 1 ton of bombs.

'Results of the above attacks varied from "unobserved" and "fair" to "excellent".'

It appears to be confirmed from Luftwaffe records that Cap d'Antifer was still operating, and passing on data to German batteries in August, 1944, a stout effort.

On the other hand, had the Allies wanted to take out Cap d'Antifer Radar station, they would have done so; they did not because the station was one of the star targets of Dr (now Sir Robert) Cockburn in his spoofing programme which was one of the main successes of the D-Day project. One of Cockburn's 'Ghost Fleets' (the one called TAXABLE) was indeed aimed directly at Cap d'Antifer so that that station's reports would suggest a powerful Allied assault fleet fronting it. For an excellent account of the D-Day spoofing, so ingenious, so versatile, so complete, see Price, *Instruments of Darkness*, Chapter 9, pp. 199–211.

2. Jones, letter to the author.

## 22

# Adieux

*1942*

Patrols scoured the roads, fields, woods, and villages. Farmyards were searched. Women were questioned as they walked out to milk in the fields, carrying their stools and pails. The French police received orders to assist in the capture of any 'commandos' who might be at large or of any suspicious persons. It was possible that the British had used the raid to land agents whose targets might be elsewhere. At eight in the morning a General's Mercedes drew up in front of the *Hotel Beauminet*, whose yard quickly filled with motor cyclists in steel helmets and a squad of the Feldgendarmerie. In the hotel office a parachutist, wounded in one arm, was brought before the General, who behaved courteously, it appeared, and spoke with the young prisoner in English. Monsieur Vennier, watching through the partly open door, saw the prisoner draw himself up and salute the General with his good arm, before he was taken out to the Feldgendarmerie escort. Three unwounded prisoners were then questioned in the guard room on the ground floor of the hotel. Madame Vennier, who understood German and some English—hence the faded *English Spoken* notice that could still be seen in the corner of the porch window—managed to pick up scraps of the interrogation. She and the maid were hidden in the partitioned serving-place of the former dining room. They admired the fresh, glowing skins of the prisoners and their stout clothing and boots. When the Germans withdrew, the two women made signs of friendship and sympathy. But the parachutists were laughing at something, and paid little attention.[1]

Two dead German soldiers lay in what had been, in better times, a private dining room used for parties or for romantic assignations. They lay for forty-eight hours under the ping-pong table, each hidden by a blanket, only his feet showing and his hands in their grey-green gloves.

Several times that day the hotel was searched from cellars to attics. At midday the meal was ready as usual for the thirty soldiers billeted there. Only seven or eight turned up. 'Where are my clients?' the little maid cried.

'Drinking cups of tea, I wouldn't be surprised,' Vennier answered. He and his wife had been awake all night, listening to the shooting, the rushing of boots, and the ringing telephone. They had thought from the amount of firing as the landing craft foamed away from the beach and the heavy German vehicles came pouring down the La Poterie road almost nose to tail that the Tommies were forcing a landing, and that soon the war might be over. Dawn brought more sober realisations. And the telephone never stopped shrilling. The Sergeant who commanded at Bruneval was still unshaven in the late afternoon, his eyes mad. Usually he was an easy fellow to get on with. Very correct.

Three days later, Madame Vennier was alone in front of the hotel, thinking what a wonderful show of geraniums they used to have there before the war, when phssst! a black well-polished front-wheel-drive Citroën drew up close beside her skirt and four men got out. Gestapo was written all over them.

'Where is the Englishman?' one of them asked her.

'Englishman! We have nothing as exotic as that round here.' It looked as though somebody, some local enemy, had lodged a denunciation. It could be very serious.

'You are hiding an English soldier.'

'How could we, when the hotel is packed like a box of dates with German soldiers?'

'You are of English origin.'

'On the contrary, until my marriage I was Swiss, and now I am French, one hundred per cent.'

'And why, in your opinion, did the English attack at Bruneval?'

'To prove to the German High Command that if they can get a footing in a place like this they can get a footing anywhere,' answered Monsieur Vennier, who had hurried, breathless, round the corner of the yard.

'Please explain yourself.'

'I am an officer of the Reserve,' said Vennier. 'It must be obvious that the English want to induce you Germans to weaken the Russian front by bringing more and yet more of your best troops back to France.'

Madame Vennier thought her husband's bold manner and the foolishness of his answers saved them from prison, because the Gestapo men deduced that they knew nothing of the Radar station up above them near the cliffs.

'You possess a wireless set?'

'Yes. In the office.'

He switched it on, but only got static. To detune it after each listening to the BBC news in French had become second nature. 'Can you certify that none of the people in Bruneval have English sympathies?'

'Certainly not! How should we be aware of their sympathies? We don't parade emotions round here. We're too busy trying to keep body and soul together on the rations.'

A day or two later a German Lieutenant requisitioned a bedroom and the maid had to sleep out, in one of the farms. He seldom slept there, but used his room, it appeared to the Venniers, as an excuse to arrive in the hotel at unusual hours of the day or the night.

One day towards the end of April, Charlemagne turned up with a couple of business friends from Le Havre. They ordered coffee and calvados, and Vennier, making up his mind not to ask for payment, even if they consumed the whole bottle, contrived to draw Charlemagne into the hall, and from there into the office. He switched on the wireless, got some music, and turned the volume full on. Then he gripped the other by the arm, and whispered fiercely right into his ear, 'It was you! It was you, Monsieur Chauveau, you and that dark man, who went down to see if the beach was mined . . . It was you or that other who planned the . . .'

'Are you insane, my friend?' Charlemagne interrupted. He suddenly looked smaller, woebegone, tormented. 'As for my friend . . .'

'What's happened?'

* * *

Exactly a month after writing his message of congratulations to Pol in a private sitting room of the Waldorf Hotel in Aldwych, Rémy returned to Paris, and that first night he drank a glass with Pol, and heard his news. The very next day Pol, as the result of a denunciation, was arrested by the Germans and imprisoned in Fresnes. Confrérie Notre Dame kept close watch and believed that the airman would be released, even if they held him for a considerable time. He had always been discreet in his secret work, and his alibis appeared to be sound.

But on Friday, May 19, Bob, Rémy's personal 'radio', was taken by the Funkabwehr. As he had always been over-bold, audacity itself—Rémy had constantly tried to make him mend his ways—there was little hope for poor Bob, or for his brother Pierre (code-name Boulot), who was soon arrested. Bob was speedily put to the torture. (He was to die of his sufferings nearly a year later, in Fresnes.) Rare indeed was the man who could withstand that type of questioning. Those who risked such mistreatment were counselled to hold out for forty-eight hours to give their friends time to cover up, and then only should they appear to give in, pitting such wits as might still be available to them against those of the inquisitors. Bob, bravely, did that. And seeking to give away secrets that were long out of date, he revealed the workings of Rémy's 'Raymond-B' code. Bob knew that Rémy had abandoned that code. But he forgot, or never knew, that the extremely efficient monitoring service of the Funkabwehr filed every message that it managed to intercept.

As soon as Bob told his interrogators that the key to Raymond-B was the *Petit Larousse Illustré*, the Funkabwehr went through their files. They knew that another prisoner's real name was Roger Dumont, and that his alias was Pol. When they came to Rémy's congratulatory message, A PACO POUR POL, the prisoner was doomed.

A year after his imprisonment at Fresnes, Pol was killed by a firing squad at Mont Valérien. The German almoner was with him when, an hour before his execution, he wrote to his family, 'All that I have done I have done as a Frenchman. I regret nothing.'

* * *

On a bitterly cold morning at the beginning of 1944 Field-Marshal Rommel came to Bruneval as part of a major tour of inspection and reappraisement, he being the officer in charge of the English Channel defences. A colonel asked the maid in the *Beauminet* to bring out two bottles of brandy for the Field-Marshal's staff. There were discussions in the roadway outside the hotel. The officers' breath hung round their heads in clouds. None of them smoked, the maid noticed; Frenchmen seemed to smoke more than Germans, these days. She was paid for the brandy, and stood with the tray of glasses, watching the smart officers in their beautiful cars sliding down to the beach. It was one of those days when the air seemed a little warmer beside the sea. They spent some time under the cliffs standing on round pebbles that had been chilled by a below-zero night. To the north, in the direction of England, the Channel was a steely, relentless grey. A depressing day.

And the following day an order came through to Bruneval for the evacuation of all civilians. On February 15 the Venniers left their *Hotel Beauminet*, where they had known prosperity and much happiness, never to return.

There is no hotel in Bruneval today. There is no LONE HOUSE up above. The Germans razed it. Spotted cattle belonging to the Le Presbytère farms shelter from sun or rain in some of the ruined outhouses. They have to watch their footing because of crumbling airshafts in the grass, rising from underground workings, dug by forced labour. There are more casemates than there were during John Frost's first visit with 'C' Company.

* * *

For eighteen months before the Bruneval affair R. V. Jones had been pressing for a complete move of the Telecommunications Research Establishment, whose physical size was now beginning to approach its great importance in the allied war effort, since all the new British Radar was under development there. 'I believed', Jones says,[2] 'that it would be easy for the Germans to detect our new transmissions from a listening post near Cherbourg, and that we therefore ought to move TRE to a site well beyond the range of radio interception. However, Swanage was a pleasant place, and the workers did not want to move. . . . After the Bruneval Raid, however, speculation started about whether the Germans might organise a reprisal; if so, Swanage was a very obvious target.'

It was indeed. TRE, like the *Würzburg*, was on the Channel's very edge, and TRE's position offered a choice of excellent beaches that could be used either for assault or for escape.

To the Establishment itself the notion of any possible move was cataclysmic. As it had grown round Worth Matravers, then expanded to Swanage, then Christchurch, then Hurn near Bournemouth, it had drawn in, and trained, its own, now highly skilled, labour. Because the staff were Dorset people, locally recruited, and because it was a holiday area (in peacetime) there was no housing problem, an important consideration in any secret establishment.

'Then came a bombshell,' A. P. Rowe says. 'There were, we were told, seventeen trainloads of German parachute troops on the other side of the English Channel preparing to attack TRE. The Prime Minister said we must leave the South Coast before the next full moon. A whole regiment of infantry arrived to protect us. They blocked the road approaches, they encircled us with barbed wire, they put demolition charges in our secret equipment, and they made our lives a misery. My own time was spent in discussions as to whether we should die to the last scientist, or run. These events made us co-operate in the task of finding a place where we could get on with the war in peace.'

When R. V. Jones had made sure that they had been informed 'through other channels' at Swanage of the arrival of a German parachute unit at Cherbourg, he went down himself to Dorset with one of his uniformed officers. 'We said nothing about the possibility of a reprisal, but we contrived to seem ill at ease throughout the visit, and our conspicuously worn pistols' (Jones was a noted marksman with a pistol) 'and steel hats testified to our apprehension.'[3]

Malvern College, the school on the side of the Malvern hills overlooking the Vale of Evesham and part of the Severn Valley, was chosen for TRE's new setting. The perimeter was made secure to military and scientific standards, new laboratories and workshops were built, power and telephones were brought in, and a nearby aerodrome was enlarged and improved. It was a splendid new base. Great things were to be accomplished there; the triumphant ghost fleets using Window that made D-Day possible were run from there; King George VI and his Queen visited Rowe and his scientists there. But Worth Matravers and the pale, pure textures of sky, sea, cliff, and grass of the Dorset coast would ever be regretted by the Radar men. When they were forced to leave Dorset it seemed that they must leave too much behind them.

'On May 25, 1942,' Rowe says,[4] 'the TRE people began to move to Malvern by train and in cars mostly so old that it seemed impossible that they could reach their destination. For the last time, I made my auto-cycle journey from Swanage to unforgettable Corfe [Castle] where, at the Old Tea House, I had lived for two years . . .'

### NOTES TO CHAPTER 22

1. Rémy, *Bruneval*, pp. 227–237, and personal investigations at Bruneval and its neighbourhood in 1970.
2. Jones, letter to the author.
3. Jones, *Minerva*, Vol. X, No. 3, p. 243.
4. Rowe, *One Story of Radar*, p. 134.

# *Bibliography*

AIR MINISTRY, The, *By Air to Battle* (HMSO, 1945), *The Rise and Fall of the German Air Force* (HMSO, 1946), *The Origins and Development of Operational Research in the Royal Air Force* (HMSO, 1963)

BABINGTON SMITH, Constance, *Evidence in Camera* (Chatto & Windus, 1958)

BARKER, Ralph (as told to), *Aviator Extraordinary*, The Sidney Cotton Story (Chatto & Windus, 1969)

BIRKENHEAD, The Earl of, *The Prof in Two Worlds* (Collins, 1961)

BLACKETT, Lord (Professor P. M. S.), *Studies of War* (Oliver & Boyd, 1962)

BRYANT, Sir Arthur, *The Turn of the Tide* (Collins, 1957)

CHURCHILL, Winston S., *The Second World War* (Cassell, 1948–54)

CLARK, Ronald W., *The Rise of the Boffins* (Phoenix House, 1962), *Tizard* (Methuen, 1965), *Sir Edward Appleton* (Pergaman Press, 1971)

COLLIER, Basil, *The Defence of the United Kingdom* (HMSO, 1957)

COWBURN, Benjamin, *No Cloak, No Dagger* (Jarrolds, 1960)

CROWTHER, J. G., and WHIDDINGTON, R., *Science at War* (HMSO, 1947)

FERGUSSON, Bernard, *The Watery Maze*, The Story of Combined Operations (Collins, 1961)

FOOT, M. R. D., *SOE in France* (HMSO, 1966)

GALLAND, Adolf, *The First and the Last* (Methuen, 1955)

GAULLE, Charles de, *Mémoires de Guerre* (Plon, 1954–9)

GOUDSMIT, Samuel A., *ALSOS: The Failure in German Science* (Sigma Books, 1947)

HARRIS, Sir Arthur, *Bomber Offensive* (Collins, 1947)

HARROD, Roy, *The Prof* (Macmillan, 1959)

HARTCUP, Guy, *The Challenge of War* (David & Charles, 1970)

HILL, Professor A. V., *The Ethical Dilemma of Science* (OUP, 1960)

*History of the Second Battalion, The Parachute Regiment* (Gale & Polden, 1946)

HORAN, Rear-Admiral H. E., *Raid on Bruneval* (THE NAVY, Journal of the Navy League, Vol. LVI, No. 3, 1951)

HUTCHINSON, Walter (ed.), *Raid on German Radio-Location Post* (Hutchinson's Pictorial History of the War, No. 10, Series 15, 1942)

INFORMATION, Ministry of, *Combined Operations 1940–1942* (HMSO, 1943), *By Air to Battle* (HMSO, 1945)

JONES, Professor R. V., *Scientific Intelligence* (Journal of the Royal United Services Institution, 1947), *Winston Leonard Spencer Churchill 1874–1965*

(Biographical Memoirs of Fellows of the Royal Society, Vol. 12, 1966), *Temptations and Risks of the Scientific Observer* (Minerva, Vol. X, No. 3, 1972)

LEVERKUEHN, P., *German Military Intelligence* (Weidenfeld & Nicolson, 1954)

LIVRY-LEVEL, Colonel Philippe, *Missions dans la R.A.F.* (Ozanne, Caen, 1951), and with Rémy (see below) *The Gates Burst Open* (Arco, 1955)

LONGMATE, Norman, *How We Lived Then* (Hutchinson, 1971)

MIDDLETON, Drew, *The Sky Suspended* (Secker & Warburg, 1960)

MINERVA, Vol. X, see JONES, Professor R. V. (International Association for Cultural Freedom, London, 1972)

PASSY (DEWAWRIN, Colonel A.), *2e Bureau, Londres* (Solar, Monte Carlo, 1947), *10 Duke Street, Londres* (Solar, 1947), *Missions Secrètes* (Plon, 1951)

PRICE, Alfred, *Instruments of Darkness* (Kimber, 1967)

REITLINGER, Gerald, *The SS: Alibi of a Nation* (Heinemann, 1956)

REMY (RENAULT, Colonel Gilbert), *Comment meurt un reseau* (Solar, Monte Carlo, 1947), *Une affaire de trahison* (Solar, 1947), *Les mains jointes* (Solar, 1948), *The Silent Company* (Barker, 1948), *Courage and Fear* (Barker, 1950), *Portrait of a Spy* (Barker, 1955), *Ten Steps to Hope* (Barker, 1960), *Bruneval: opération coup de croc* (Editions France Empire, 1968). And with LIVRY-LEVEL (see above), *The Gates Burst Open* (Arco, 1955)

RICHARDS, Denis, *Royal Air Force 1939–1945. Vol. 1. The Fight at Odds* (HMSO, 1953)

ROWE, A. P., *One Story of Radar* (CUP, 1948)

SAUNDERS, Hilary St George, *The Red Beret* (Michael Joseph, 1950)

SPEARS, Major-General Sir Edward, *Assignment to Catastrophe* (Heinemann, 1947)

SWINTON, Lord, *I Remember* (Hutchinson, 1948)

WALLACE, Graham, *R.A.F. Biggin Hill* (Putnam, 1957)

WATSON-WATT, Sir Robert, *Three Steps to Victory* (Odham's Press, 1958)

WEBSTER, Sir Charles, and FRANKLAND, Noble, *The Strategic Air Offensive Against Germany* (HMSO, 1961)

WING-COMMANDER, A, *The Bruneval Raid* (Royal Air Force Journal, Vol. 2, No. 5, 1944)

WOOD, Derek, and DEMPSTER, Derek, *The Narrow Margin* (Hutchinson, 1961)

# *Abbreviations*

ABDA American/British Defence Association
AEAF Allied Expeditionary Air Forces
AOC Aircraft Operating Company ('Lemnos' Hemming)
ASV Air to Surface Vessel (Radar)
BCRA(M) Bureau Central de Renseignments et d'Action (Militaire)
CAS Chief of Air Staff
CH Chain Home (Radar)
CHL Chain Home Low (Radar)
CID Committee of Imperial Defence
CND Confrérie Notre Dame
DEM Détection Electro Magnétique
Do Dornier (bomber)
D/T Dezimeter Telegraphie (Radar)
DTN Defence Teleprinter Network
DZ Dropping Zone
FMG Funk Messgerät (Radar)
FUP Forming up Point
GCI Ground Controlled Interception (Radar)
GEC General Electric Company
GEMA German Radar Company
GFP Geheime Feldpolizei
GPO General Post Office
HE High Explosive (bombs)
$_2$S Centimetric Radar bombing aid
IF Radar Intermediate Frequency (unit)
IFF Identification Friend from Foe (Radar)
JG Jagdgeschwader, fighter wing
Ju Junker (bomber or fighter)
KG Kampfgruppe, squadron
KGr Kampfgruppe, independent squadron
LCA Assault Landing Craft
Me Messerschmitt (fighter)
MGB Motor Gunboat
NID Naval Intelligence (Department)
OKL Supreme Command, Luftwaffe
PDU Photographic Development Unit
POW Prisoner of War
PRU Photographic Reconnaissance Unit
PTS Parachute Training Squadron
RAE Royal Aircraft Establishment (Farnborough)
RDF Radio Direction Finding (Radar)
RE Royal Engineers
R/T Radio Telephony
RX Reception (Radar)
SD Sicherheitdienst
SOE Special Operations Executive
TRE Telecommunications Research Establishment
TX Transmission (Radar)
VHF Very High Frequency
VHF/DF Very High Frequency Direction Finding
WAAF Women's Auxiliary Air Force

# Index